Canadian Mathematical Society
Société mathématique du Canada

T0255080

For other titles published in this series, go to
www.springer.com/series/4318

Károly Bezdek

Classical Topics in Discrete Geometry

 Springer

Károly Bezdek
Department of Mathematics and Statistics
University of Calgary
2500 University Drive NW
Calgary, Alberta, T2N 1N4
Canada
bezdek@math.ucalgary.ca

Editors-in-Chief
Rédacteurs-en-chef
K. Dilcher
K. Taylor
Department of Mathematics and Statistics
Dalhousie University
Halifax, Nova Scotia, B3H 3J5
Canada
cbs-editors@cms.math.ca

The author was supported by the Canada Research Chair program as well as a
Natural Sciences and Engineering Research Council of Canada Discovery Grant.

ISBN 978-1-4614-2620-2   ISBN 978-1-4419-0600-7 (eBook)
DOI 10.1007/978-1-4419-0600-7
Springer New York Dordrecht Heidelberg London

Mathematics Subject Classification (2010): 52A38, 52A40, 52B60, 52C17, 52C22

Springer is part of Springer Science+Business Media (www.springer.com)

To my wife Éva and our sons Dániel, Máté, and Márk

Preface

Geometry is a classical core part of mathematics which, with its birth, marked the beginning of the mathematical sciences. Thus, not surprisingly, geometry has played a key role in many important developments of mathematics in the past, as well as in present times. While focusing on modern mathematics, one has to emphasize the increasing role of discrete mathematics, or equivalently, the broad movement to establish discrete analogues of major components of mathematics. In this way, the works of a number of outstanding mathematicians including H.S. M. Coxeter (Canada), C. A. Rogers (United Kingdom), and L. Fejes-Tóth (Hungary) led to the new and fast developing field called discrete geometry. One can briefly describe this branch of geometry as the study of discrete arrangements of geometric objects in Euclidean, as well as in non-Euclidean spaces. This, as a classical core part, also includes the theory of polytopes and tilings in addition to the theory of packing and covering. Discrete geometry is driven by problems often featuring a very clear visual and applied character. The solutions use a variety of methods of modern mathematics, including convex and combinatorial geometry, coding theory, calculus of variations, differential geometry, group theory, and topology, as well as geometric analysis and number theory.

The present book is centered around topics such as sphere packings, packings by translates of convex bodies, coverings by homothetic bodies, illumination and related topics, coverings by planks and cylinders, monotonicity of the volume of finite arrangements of spheres, and ball-polyhedra. The first part of the book gives an overview of the most relevant state-of-the-art research problems, including the problem of finding densest sphere packings, estimating the (surface) volume of Voronoi cells in sphere packings, studying the Boltyanski–Hadwiger conjecture, the affine invariant version of Tarski's plank problem, and the Kneser–Poulsen conjecture, just to mention a few classical ones. The second part of the book is a collection of selected proofs that have been discovered in the last ten years, and have not yet appeared in any book or monograph. I have made definite efforts to structure those proofs such that they are presentable within a normal lecture (resp., seminar). The majority of the proofs presented in the second part of the book have been developed by the author, or were results of the author's joint work with a number of colleagues. It is a particular pleasure for me to acknowledge my long-lasting collaboration with Bob Connelly (Cornell University, USA) and my brother András Bezdek (Auburn University, USA and Rényi Institute, Hungary).

This book is aimed at advanced undergraduate and early graduate students, as well as interested researchers. In addition to leading the reader to the frontiers of geometric research, my work gives a short introduction to the classical cores of discrete geometry. The forty-some research problems listed are intended to encourage further research. The following three books on related topics might provide good supplemental material: J. Matousek, *Lectures on Discrete Geometry*, Springer, 2002; Ch. Zong, *The Cube: A Window to Convex and Discrete Geometry*, Cambridge Univ. Press., 2006 and P. Gruber, *Convex and Discrete Geometry*, Springer, 2007.

The idea of writing this book came to me while I was preparing my lecture at the COE Workshop on Sphere Packings at Kyushu University (Fukuoka, Japan) in November 2004. The final version of the manuscript was prepared during my visit to Eötvös University (Budapest, Hungary) in the summer of 2009 and in the fall of 2009, while on research leave from the University of Calgary.

The present book is based on the material of some graduate-level courses and research seminar lectures I gave between 2005-2009 at the Department of Mathematics and Statistics of the University of Calgary. I am very much indebted to all my students and colleagues who attended my lectures and actively participated in the discussions. It gave me great satisfaction that those lectures motivated a good deal of further research, and, in fact, some of the results in this book were obtained by the participants during and after my lectures. My gratitude is due to my colleagues T. Bisztriczky, Richard Guy, Ferenc Fodor, Joseph Ling, Deborah Oliveros-Braniff, Jonathan Schaer, Dihn Thi, Csaba D. Tóth, Yuriy Zinchenko at the University of Calgary, and K. Böröczky, B. Csikós, A. Heppes, Gy. Kiss at Eötvös University and my former graduate students Zsolt Lángi, Wesley Maciejewski, Márton Naszódi, Bouchra Mika Sabbagh, Peter Papez, Máté Salát at the University of Calgary. Also, I want to thank the contributions of my three sons, Dániel, Máté, and Márk to some of the topics in this book. Last but not least I wish to thank my wife, Éva, whose strong support and encouragement helped me a great deal during the long hours of writing.

January 2010, Calgary   Károly Bezdek, Canada Research Chair

Contents

Part II Selected Proofs

Part I

Classical Topics Revisited

1

Sphere Packings

1.1 Kissing Numbers of Spheres

The main problem in this section is fondly known as the kissing number problem. The *kissing number* τ_d is the maximum number of nonoverlapping d-dimensional balls of equal size that can touch a congruent one in the d-dimensional Euclidean space $\mathbb{E}^d$. In three dimensions this question was the subject of a famous discussion between Isaac Newton and David Gregory in 1694. So, it is not surprising that the literature on the kissing number problem is an extensive one. Perhaps the best source of information on this problem is the book [108] by Conway and Sloane. In what follows we give a short description of the present status of this problem.

$\tau_2 = 6$ is trivial. However, determining the value of τ_3 is not a trivial issue. Actually the first complete and correct proof of $\tau_3 = 12$ was given by Schütte and van der Waerden [229] in 1953. The often-cited proof of Leech [189], which is impressively short, contrary to the common belief does contain some gaps. It can be completed though; see, for example, [192] as well as the more recent paper [193]. Further recent proofs can be found in [92], [8], and in [205]. (For additional information on all this see the very visual paper [100].) Thus, we have the following theorem.

Theorem 1.1.1 $\tau_3 = 12$.

Following the chronological ordering, here are the major inputs on the kissing numbers of Euclidean balls of dimension larger than 3. Coxeter [111] conjectured and Böröczky [89] proved the theorem stated as Theorem 1.1.2, where $F_d(\alpha) := \frac{2^d T_{d-1}(2\alpha)}{(d+1)!\omega_{d+1}}$ is the Schläfli function with $T_{d-1}(2\alpha)$ standing for the spherical volume of a regular spherical $(d-1)$-dimensional simplex of dihedral angle 2α and with ω_{d+1} denoting the volume of a $(d+1)$-dimensional unit ball in $\mathbb{E}^{d+1}$.

Theorem 1.1.2 $\tau_d \leq \frac{2F_{d-1}(\beta)}{F_d(\beta)}$, where $\beta := \frac{1}{2}\mathrm{arcsec}(d)$.

K. Bezdek, *Classical Topics in Discrete Geometry*, CMS Books in Mathematics,
DOI 10.1007/978-1-4419-0600-7_1, © Springer Science+Business Media, LLC 2010

It was a breakthrough when Delsarte's linear programming method (for details see, for example, [214]) was applied to the kissing number problem and also, when Kabatiansky and Levenshtein [173] succeeded in improving the upper bound of the previous theorem for large d as follows. The lower bound mentioned below was found by Wyner [248] several years earlier.

Theorem 1.1.3 $2^{0.2075d(1+o(1))} \leq \tau_d \leq 2^{0.401d(1+o(1))}$.

Unfortunately, the gap between the lower and upper bounds is exponential. Still, Levenshtein [190] and Odlyzko and Sloane [211], independently of each other, were able to prove the following exact values for τ_d.

Theorem 1.1.4 $\tau_8 = 240$ and $\tau_{24} = 196560$.

In addition, Bannai and Sloane [19] proved the following uniqueness result.

Theorem 1.1.5 *There is a unique way (up to isometry) of arranging* 240 *(resp.,* 196560*) non-overlapping unit spheres in 8-dimensional (resp., 24-dimensional) Euclidean space such that they touch another unit sphere.*

More recently Musin [203], [204] extending Delsarte's method found the kissing number of 4-dimensional Euclidean balls. Thus, we have

Theorem 1.1.6 $\tau_4 = 24$.

The following is generally believed to be true, but so far no one has been able to prove it.

Conjecture 1.1.7 *There is a unique way (up to isometry) of arranging* 24 *non-overlapping unit spheres in 4-dimensional Euclidean space such that they touch another unit sphere.*

It was mentioned by the author in [62] that basic rigidity techniques (such as the ones discussed in [222]) imply the following theorem that one can view as the local version of Conjecture 1.1.7. (As a next step towards a proof of Conjecture 1.1.7, it would be helpful to find a proper explicit value for ϵ in the statement below.)

Theorem 1.1.8 *Take a unit ball* $\mathbf{B}$ *of* $\mathbb{E}^4$ *touched by 24 other (non-overlapping) unit balls* $\mathbf{B}_1, \mathbf{B}_2, \ldots, \mathbf{B}_{24}$ *with centers* $\mathbf{c}_1, \mathbf{c}_2, \ldots, \mathbf{c}_{24}$ *such that the centers* $\mathbf{c}_1, \mathbf{c}_2, \ldots, \mathbf{c}_{24}$ *form the vertices of a regular 24-cell* $\{3, 4, 3\}$ *in* $\mathbb{E}^4$. *Then there exists an* $\epsilon > 0$ *with the following property: if the non-overlapping unit balls* $\mathbf{B}'_1, \mathbf{B}'_2, \ldots, \mathbf{B}'_{24}$ *with centers* $\mathbf{c}'_1, \mathbf{c}'_2, \ldots, \mathbf{c}'_{24}$ *are chosen such that* $\mathbf{B}'_1, \mathbf{B}'_2, \ldots, \mathbf{B}'_{24}$ *are all tangent to* $\mathbf{B}$ *in* $\mathbb{E}^4$ *and for each* $i, 1 \leq i \leq 24$ *the Euclidean distance between* $\mathbf{c}_i$ *and* $\mathbf{c}'_i$ *is at most* ϵ, *then* $\mathbf{c}'_1, \mathbf{c}'_2, \ldots, \mathbf{c}'_{24}$ *form the vertices of a regular 24-cell* $\{3, 4, 3\}$ *in* $\mathbb{E}^4$.

For the best known upper bounds on the kissing number for dimensions up to 24 (based on semidefinite programming) we refer the interested reader to [9] and [198]. Also, we note that there is a long list of highest kissing numbers presently known in dimensions from 32 to 128 and published in [210]. Last but not least the paper [123] of Edel, Rains, and Sloane describes some elementary and amazingly efficient constructions.

1.2 One-Sided Kissing Numbers of Spheres

The *one-sided kissing number* $B(d)$ of a d-dimensional ball say, $\mathbf{B}$ in $\mathbb{E}^d$ is the largest number of non-overlapping translates of $\mathbf{B}$ that touch $\mathbf{B}$ and that all lie in a closed supporting halfspace of $\mathbf{B}$. The term "one-sided kissing number" was proposed by the author in [62] and the notation $B(d)$ was introduced by Musin in [206]. It is obvious that the one-sided kissing number of any circular disk in $\mathbb{E}^2$ is 4; that is, $B(2) = 4$. However, the 3-dimensional analogue statement is harder to come up with. Actually, this problem was raised by L. Fejes Tóth and H. Sachs in [139]. As it turns out, the one-sided kissing number of a 3-dimensional Euclidean ball is 9. This was first proved in [130] (see also [225] and [31] for other proofs).

Theorem 1.2.1 *The one-sided kissing number of the 3-dimensional Euclidean ball is 9; that is, $B(3) = 9$.*

In fact, we know a bit more; namely, it is proved in [178] that in the case of the 3-dimensional Euclidean ball the maximal one-sided kissing arrangement is unique up to isometry. As we have mentioned before, Musin [204] has recently published a proof of the long-standing conjecture that the kissing number of the 4-dimensional Euclidean ball is 24. Based on that he [206] has given a proof of the following related statement that has been conjectured by the author [62].

Theorem 1.2.2 *The one-sided kissing number of the 4-dimensional Euclidean ball is 18; that is, $B(4) = 18$.*

Very recently, using semidefinite programming, Bachoc and Vallentin [10] were able to prove the following theorem.

Theorem 1.2.3 *The one-sided kissing number of the 8-dimensional Euclidean ball is 183; that is, $B(8) = 183$.*

We close this section with the relevant challenging conjectures of Musin [206] (see also [21]).

Conjecture 1.2.4 $B(5) = 32$ *and* $B(24) = 144855$.

1.3 On the Contact Numbers of Finite Sphere Packings

Let $\mathbf{B}$ be a ball in $\mathbb{E}^d$. Then the contact graph of an arbitrary finite packing by non-overlapping translates of $\mathbf{B}$ in $\mathbb{E}^d$ is the (simple) graph whose vertices correspond to the packing elements and whose two vertices are connected by an edge if and only if the corresponding two packing elements touch each other. One of the most basic questions on contact graphs is to find the maximum number of edges that a contact graph of n non-overlapping translates of the given Euclidean ball $\mathbf{B}$ can have in $\mathbb{E}^d$. Harborth [166] proved the following remarkable result on the contact graphs of congruent circular disk packings in $\mathbb{E}^2$.

Theorem 1.3.1 *The maximum number of touching pairs in a packing of n congruent circular disks in $\mathbb{E}^2$ is precisely*

$$\lfloor 3n - \sqrt{12n - 3} \rfloor.$$

The analogue question in the hyperbolic plane has been studied by Bowen in [86]. We prefer to quote his result in the following geometric way.

Theorem 1.3.2 *Consider circle packings in the hyperbolic plane, by finitely many congruent circles, which maximize the number of touching pairs for the given number of congruent circles. Then such a packing must have all of its centers located on the vertices of a triangulation of the hyperbolic plane by congruent equilateral triangles, provided the diameter D of the circles is such that an equilateral triangle in the hyperbolic plane of side length D has each of its angles equal to $\frac{2\pi}{N}$ for some $N > 6$.*

It is not hard to see that one can extend the above result to $\mathbb{S}^2$ exactly in the way as the above phrasing suggests. However, we get a more general approach if we do the following. Take n non-overlapping unit diameter balls in a convex position in $\mathbb{E}^3$; that is, assume there exists a 3-dimensional convex polyhedron whose vertices are center points. Moreover, each center point belongs to the boundary of that convex polyhedron, where $n \geq 4$ is a given integer. Obviously, the shortest distance among the center points is at least one. Then count the unit distances showing up between pairs of center points but count only those pairs that generate a unit line segment on the boundary of the given 3-dimensional convex polyhedron. Finally, maximize this number for the given n and label this maximum by $c(n)$. In the following statement of D. Bezdek [36] the convex polyhedra entering are called "generalized deltahedra" (or in short, "g-deltahedra") (see also [37]) mainly because that family of convex polyhedra includes all "deltahedra" classified quite some time ago by Freudenthal and van der Waerden in [144].

Theorem 1.3.3 $c(n) \leq 3n - 6$, *where equality is attained for infinitely many n namely, for those for which there exists a 3-dimensional convex polyhedron each face of which is an edge-to-edge union of some regular triangles of side* *length one such that the total number of generating regular triangles on the boundary of the convex polyhedron is precisely $2n - 4$ with a total number of $3n - 6$ sides of length one and with a total number of n vertices.*

Theorem 1.3.3 proposes to find a proper classification for g-deltahedra, a question that is still open. Some partial results on that can be found in the recent paper [37] of D. Bezdek, the main result of which states that the regular icosahedron has the smallest isoperimetric quotient among all g-deltahedra. For the sake of completeness we mention that this result supports the still open *Icosahedral Conjecture* of Steiner (1841), according to which among all convex polyhedra isomorphic to an icosahedron (i.e. having the same face structure as an icosahedron) the regular icosahedron has the smallest isoperimetric quotient. Another interesting result on g-deltahedra was obtained by D. Bezdek in [36]. It claims that every g-deltahedron has an edge unfolding. This is part of the general problem, raised by Shephard (1975) and motivated also by some drawings of Dürer (1525), of whether every convex polyhedron has an edge unfolding, that is can be cut along some of its edges and then folded into a single planar polygon without overlap. For more details on this we refer the interested reader to the lavishly illustrated book [121] of Demaine and O'Rourke.

Now, we are ready to phrase the *Contact Number Problem* of finite congruent sphere packings in $\mathbb{E}^3$. For a given positive integer $n \geq 2$ find the largest number $C(n)$ of touching pairs in a packing of n congruent balls in $\mathbb{E}^3$. One can regard this problem as a combinatorial relative of the Kepler conjecture on the densest unit sphere packings in $\mathbb{E}^3$. It is easy to see that $C(2) = 1, C(3) = 3, C(4) = 6, C(5) = 9, C(6) = 12, C(7) = 15, C(8) = 18$ and $C(9) = 21$. So, it is natural to continue with the following question.

Problem 1.3.4 *Prove or disprove that $C(10) = 24, C(11) = 28, C(12) = 32$ and $C(13) = 36$. In general, prove or disprove that $C(n)$ can be achieved in a packing of n unit balls in $\mathbb{E}^3$ consisting of parallel layers of unit balls each being a subset of the densest infinite hexagonal layer of unit balls.*

For a general n it seems challenging enough to search for good (lower and) upper bounds for $C(n)$. In connection with this problem, the author [62] has proved the following estimate.

Theorem 1.3.5 $C(n) < 6n - 0.59n^{\frac{2}{3}}$ *for all $n \geq 2$.*

1.4 Lower Bounds for the (Surface) Volume of Voronoi Cells in Sphere Packings

Recall that a family of non-overlapping 3-dimensional balls of radii 1 in Euclidean 3-space, $\mathbb{E}^3$ is called a unit ball packing in $\mathbb{E}^3$. The density of the packing is the proportion of space covered by these unit balls. The sphere packing problem asks for the densest packing of unit balls in $\mathbb{E}^3$. The conjecture that the density of any unit ball packing in $\mathbb{E}^3$ is at most $\frac{\pi}{\sqrt{18}} = 0.74078\ldots$ is often attributed to Kepler's statement of 1611. The problem of proving the Kepler conjecture appears as part of Hilbert's 18th problem [168]. Using an ingenious argument which works in any dimension, Rogers [218] obtained the upper bound $0.77963\ldots$ for the density of unit ball packings in $\mathbb{E}^3$. This bound has been improved by Lindsey [191], and Muder [201], [202] to $0.773055\ldots$. Hsiang [170], [171] proposed an elaborate line of attack (along the ideas of L. Fejes Tóth suggested 40 years earlier), but his claim that he settled Kepler's conjecture seems exaggerated. However, so far no one has found any gap in the approach of Hales [156], [157], [158], [159], [160], [161], [162], [141], [163] although no one has been able to fully verify it either. This is not too surprising, given that the detailed argument is described in several papers and relies on long computer-aided calculations of more than 5000 subproblems. Thus, after several years of extremely hard work, Hales with the help of Ferguson [141] has been able to finish his complex project and they were able to come up with the most detailed and complete-looking computer-supported proof of the longstanding conjecture of Kepler that is presently known. We summarize their heroic achievement in a short statement.

Theorem 1.4.1 *The densest packing of unit balls in* $\mathbb{E}^3$ *has density* $\frac{\pi}{\sqrt{18}}$, *which is attained by the "cannonball packing".*

For several of the above-mentioned papers Voronoi cells of unit ball packings play a central role. Recall that the Voronoi cell of a unit ball in a packing of unit balls in $\mathbb{E}^3$ is the set of points that are not farther away from the center of the given ball than from any other ball's center. As is well known, the Voronoi cells of a unit ball packing in $\mathbb{E}^3$ form a tiling of $\mathbb{E}^3$. One of the most attractive problems on Voronoi cells is the *Dodecahedral Conjecture* of L. Fejes Tóth published in [132]. According to this the volume of any Voronoi cell in a packing of unit balls in $\mathbb{E}^3$ is at least as large as the volume of a regular dodecahedron with inradius 1. Very recently Hales and McLaughlin [164], [165] have succeeded in proving this long-standing conjecture of L. Fejes Tóth. Thus, we have the following theorem.

Theorem 1.4.2 *The volume of any Voronoi cell in a packing of unit balls in* $\mathbb{E}^3$ *is at least as large as the volume of a regular dodecahedron with inradius* 1.

We wish to mention that although neither Theorem 1.4.1 nor Theorem 1.4.2 implies the other, their proofs follow a similar outline and share a significant number of methods, and in particular, both are based on long computer calculations.

As a next step towards a better understanding of the underlying geometry, it seems natural to investigate the following strengthened version of the Dodecahedral Conjecture, which we call the *Strong Dodecahedral Conjecture*. It was first articulated in [55].

Conjecture 1.4.3 *The surface area of any Voronoi cell in a packing with unit balls in* $\mathbb{E}^3$ *is at least as large as* 16.6508..., *the surface area of a regular dodecahedron of inradius* 1.

It is easy to see that Conjecture 1.4.3 implies Theorem 1.4.2. The first efforts for a proof of the Strong Dodecahedral Conjecture were made by the author and Daróczy-Kiss [63]. In order to phrase their result properly we need to introduce a bit of terminology. A face cone of a Voronoi cell in a packing with unit balls in $\mathbb{E}^3$ is the convex hull of the chosen face and the center of the unit ball sitting in the given Voronoi cell. The surface area density of a unit ball in a face cone is simply the spherical area of the region of the unit sphere (centered at the apex of the face cone) that belongs to the face cone divided by the Euclidean area of the face. It should be clear from these definitions that if we have an upper bound for the surface area density in face cones of Voronoi cells, then the reciprocal of this upper bound times 4π (the surface area of a unit ball) is a lower bound for the surface area of Voronoi cells. Now we are ready to state the main theorem of [63].

Theorem 1.4.4 *The surface area density of a unit ball in any face cone of a Voronoi cell in an arbitrary packing of unit balls of* $\mathbb{E}^3$ *is at most*

$$\frac{-9\pi + 30\arccos\left(\frac{\sqrt{3}}{2}\sin\left(\frac{\pi}{5}\right)\right)}{5\tan\left(\frac{\pi}{5}\right)} = 0.77836\ldots,$$

and so the surface area of any Voronoi cell in a packing with unit balls in $\mathbb{E}^3$ *is at least*

$$\frac{20\pi\tan\left(\frac{\pi}{5}\right)}{-9\pi + 30\arccos\left(\frac{\sqrt{3}}{2}\sin\left(\frac{\pi}{5}\right)\right)} = 16.1445\ldots.$$

Moreover, the above upper bound 0.77836... *for the surface area density is best possible in the following sense. The surface area density in the face cone of any n-sided face with* $n = 4, 5$ *of a Voronoi cell in an arbitrary packing of unit balls of* $\mathbb{E}^3$ *is at most*

$$\frac{3(2-n)\pi + 6n\arccos\left(\frac{\sqrt{3}}{2}\sin(\frac{\pi}{n})\right)}{n\tan\left(\frac{\pi}{n}\right)}$$

and equality is achieved when the face is a regular n-gon inscribed in a circle of radius $\dfrac{1}{\sqrt{3}\cos\left(\frac{\pi}{n}\right)}$ *and positioned such that it is tangent to the corresponding unit ball of the packing at its center.*

The following recent improvement was obtained in [5].

Theorem 1.4.5 *The surface area of any Voronoi cell in a packing with unit balls in* $\mathbb{E}^3$ *is at least* 16.1977....

Recall that the Voronoi cell of a unit ball in a packing of unit balls in $\mathbb{E}^d$ is the set of points that are not farther away from the center of the given ball than from any other ball's center. As is well known, the Voronoi cells of a unit ball packing in $\mathbb{E}^d$ form a tiling of $\mathbb{E}^d$. One of the most attractive results on the sphere packing problem was proved by C. A. Rogers [218] in 1958. It was rediscovered by Baranovskii [20] and extended to spherical and hyperbolic spaces by Böröczky [89]. It can be phrased as follows. Take a regular d-dimensional simplex of edge length 2 in $\mathbb{E}^d$ and then draw a d-dimensional unit ball around each vertex of the simplex. Let σ_d denote the ratio of the volume of the portion of the simplex covered by balls to the volume of the simplex.

Theorem 1.4.6 *The volume of any Voronoi cell in a packing of unit balls in $\mathbb{E}^d$ is at least $\frac{\omega_d}{\sigma_d}$, where ω_d denotes the volume of a d-dimensional unit ball.*

The following strengthening of Theorem 1.4.6 has been proved by the author in [55]. (See also [54] for a somewhat simpler proof.)

Theorem 1.4.7 *The surface volume of any Voronoi cell in a packing of unit balls in $\mathbb{E}^d, d \geq 2$ is at least $\frac{d\omega_d}{\sigma_d}$.*

Indeed, Theorem 1.4.7 implies Theorem 1.4.6 by observing that the volume of a Voronoi cell in a packing of unit balls in $\mathbb{E}^d$ is at least as large as $\frac{1}{d}$ times the surface volume of the Voronoi cell in question. The next theorem due to the author [56] improves the estimate of Theorem 1.4.7 even further for all $d \geq 8$. For this we need a bit of notation. As usual, let $\lin(\cdot)$, $\aff(\cdot)$, $\conv(\cdot)$, $\vol_d(\cdot)$, ω_d, $\Svol_{d-1}(\cdot)$, $\dist(\cdot,\cdot)$, $\|\cdot\|$, and $\mathbf{o}$ refer to the linear hull, the affine hull, the convex hull in $\mathbb{E}^d$, the d-dimensional Euclidean volume measure, the d-dimensional volume of a d-dimensional unit ball, the $(d-1)$-dimensional spherical volume measure, the distance function in $\mathbb{E}^d$, the standard Euclidean norm, and to the origin in $\mathbb{E}^d$.

Let $\conv\{\mathbf{o}, \mathbf{w}_1, \ldots, \mathbf{w}_d\}$ be a d-dimensional simplex having the property that the linear hull $\lin\{\mathbf{w}_j - \mathbf{w}_i | i < j \leq d\}$ is orthogonal to the vector $\mathbf{w}_i$ in $\mathbb{E}^d, d \geq 8$ for all $1 \leq i \leq d - 1$; that is, let

$$\conv\{\mathbf{o}, \mathbf{w}_1, \ldots, \mathbf{w}_d\}$$

be a d-dimensional orthoscheme in $\mathbb{E}^d$. Moreover, let

$$\|\mathbf{w}_i\| = \sqrt{\frac{2i}{i+1}} \quad \text{for all} \ \ 1 \leq i \leq d.$$

It is clear that in the right triangle $\triangle \mathbf{w}_{d-2}\mathbf{w}_{d-1}\mathbf{w}_d$ with right angle at the vertex $\mathbf{w}_{d-1}$ we have the inequality $\|\mathbf{w}_d - \mathbf{w}_{d-1}\| = \sqrt{\frac{2}{d(d+1)}} < \sqrt{\frac{2}{(d-1)d}} = \|\mathbf{w}_{d-1} - \mathbf{w}_{d-2}\|$ and therefore $\angle \mathbf{w}_{d-1}\mathbf{w}_{d-2}\mathbf{w}_d < \frac{\pi}{4}$. Now, in the plane $\aff\{\mathbf{w}_{d-2}, \mathbf{w}_{d-1}, \mathbf{w}_d\}$ of the triangle $\triangle \mathbf{w}_{d-2}\mathbf{w}_{d-1}\mathbf{w}_d$ let

$$\vartriangleleft \mathbf{w}_{d-2}\mathbf{w}_d\mathbf{w}_{d+1}$$

denote the circular sector of central angle

$$\angle \mathbf{w}_d\mathbf{w}_{d-2}\mathbf{w}_{d+1} = \frac{\pi}{4} - \angle \mathbf{w}_{d-1}\mathbf{w}_{d-2}\mathbf{w}_d$$

and of center $\mathbf{w}_{d-2}$ sitting over the circular arc with endpoints $\mathbf{w}_d, \mathbf{w}_{d+1}$ and radius $\|\mathbf{w}_d - \mathbf{w}_{d-2}\| = \|\mathbf{w}_{d+1} - \mathbf{w}_{d-2}\|$ such that

$$\vartriangleleft \mathbf{w}_{d-2}\mathbf{w}_d\mathbf{w}_{d+1} \text{ and } \vartriangle \mathbf{w}_{d-2}\mathbf{w}_{d-1}\mathbf{w}_d$$

are adjacent along the line segment $\mathbf{w}_{d-2}\mathbf{w}_d$ and are separated by the line of $\mathbf{w}_{d-2}\mathbf{w}_d$. Then let

$$D(\mathbf{w}_{d-2}, \mathbf{w}_{d-1}, \mathbf{w}_d, \mathbf{w}_{d+1}) = \vartriangle \mathbf{w}_{d-2}\mathbf{w}_{d-1}\mathbf{w}_d \cup \vartriangleleft \mathbf{w}_{d-2}\mathbf{w}_d\mathbf{w}_{d+1}$$

be the convex domain generated by the triangle $\vartriangle \mathbf{w}_{d-2}\mathbf{w}_{d-1}\mathbf{w}_d$ with constant angle

$$\angle \mathbf{w}_{d-1}\mathbf{w}_{d-2}\mathbf{w}_{d+1} = \frac{\pi}{4}.$$

Now, let

$$\mathbf{W} = \text{conv}\big(\{\mathbf{o}, \mathbf{w}_1, \ldots, \mathbf{w}_{d-3}\} \cup D(\mathbf{w}_{d-2}, \mathbf{w}_{d-1}, \mathbf{w}_d, \mathbf{w}_{d+1})\big)$$

be the d-dimensional wedge (or cone) with $(d-1)$-dimensional base

$$Q_W = \text{conv}\big(\{\mathbf{w}_1, \ldots, \mathbf{w}_{d-3}\} \cup D(\mathbf{w}_{d-2}, \mathbf{w}_{d-1}, \mathbf{w}_d, \mathbf{w}_{d+1})\big) \text{ and apex } \mathbf{o}.$$

Finally, if $\mathbf{B} = \{\mathbf{x} \in \mathbb{E}^d | \text{ dist}(\mathbf{o}, \mathbf{x}) = \|\mathbf{x}\| \leq 1\}$ denotes the d-dimensional unit ball centered at the origin $\mathbf{o}$ of $\mathbb{E}^d$ and $\mathbb{S}^{d-1} = \{\mathbf{x} \in \mathbb{E}^d | \text{ dist}(\mathbf{o}, \mathbf{x}) = \|\mathbf{x}\| = 1\}$ denotes the $(d-1)$-dimensional unit sphere centered at $\mathbf{o}$, then let

$$\hat{\sigma}_d = \frac{\text{Svol}_{d-1}(\mathbf{W} \cap \mathbb{S}^{d-1})}{\text{vol}_{d-1}(Q_W)} = \frac{\text{vol}_d(\mathbf{W} \cap \mathbf{B})}{\text{vol}_d(\mathbf{W})}$$

be the the surface density (resp., volume density) of the unit sphere $\mathbb{S}^{d-1}$ (resp., of the unit ball $\mathbf{B}$) in the wedge W. For the sake of completeness we remark that as the regular d-dimensional simplex of edge length 2 can be dissected into $(d+1)!$ pieces each being congruent to $\text{conv}\{\mathbf{o}, \mathbf{w}_1, \ldots, \mathbf{w}_d\}$ therefore

$$\sigma_d = \frac{\text{vol}_d(\text{conv}\{\mathbf{o}, \mathbf{w}_1, \ldots, \mathbf{w}_d\} \cap \mathbf{B})}{\text{vol}_d(\text{conv}\{\mathbf{o}, \mathbf{w}_1, \ldots, \mathbf{w}_d\})}.$$

Now, we are ready to state the main result of [56]. Recall that the surface density of any unit sphere in its Voronoi cell in a unit sphere packing of $\mathbb{E}^d$ is defined as the ratio of the surface volume of the unit sphere to the surface volume of its Voronoi cell.

Theorem 1.4.8 *The surface volume of any Voronoi cell in a packing of unit balls in the d-dimensional Euclidean space $\mathbb{E}^d$, $d \geq 8$ is at least $\frac{d\omega_d}{\widehat{\sigma}_d}$; that is, the surface density of any unit sphere in its Voronoi cell in a unit sphere packing of $\mathbb{E}^d$, $d \geq 8$ is at most $\widehat{\sigma}_d$. Thus, the volume of any Voronoi cell in a packing of unit balls in $\mathbb{E}^d$, $d \geq 8$ is at least $\frac{\omega_d}{\widehat{\sigma}_d}$ and so, the (upper) density of any unit ball packing in $\mathbb{E}^d$, $d \geq 8$ is at most $\widehat{\sigma}_d$ ($< \sigma_d$).*

Last but not least we note that the proof of Theorem 1.4.8 published in [56] gives a proof of the following even stronger statement. Take a Voronoi cell of a unit ball in a packing of unit balls in the d-dimensional Euclidean space $\mathbb{E}^d$, $d \geq 8$ and then take the intersection of the given Voronoi cell with the closed d-dimensional ball of radius $\sqrt{\frac{2d}{d+1}}$ concentric to the unit ball of the Voronoi cell. Then the surface area of the truncated Voronoi cell is at least $\frac{d \cdot \omega_d}{\widehat{\sigma}_d}$.

1.5 On the Density of Sphere Packings in Spherical Containers

In this section we propose a way for investigating an analogue of Kepler's problem for finite packings of congruent balls in hyperbolic 3-space $\mathbb{H}^3$. The idea goes back to the theorems of L. Fejes Tóth [133] in $\mathbb{E}^2$, J. Molnár [199] in $\mathbb{S}^2$, and the author [39], [40] in $\mathbb{H}^2$ which, in short, can be summarized as follows.

Theorem 1.5.1 *If at least two congruent circular disks are packed in a circular disk in the plane of constant curvature, then the packing density is always less than $\frac{\pi}{\sqrt{12}}$.*

The hyperbolic case of this theorem conjectured by L. Fejes Tóth and proved by the author in [39] (see also [40]) is truly of hyperbolic nature because there are (infinite) packings of congruent circular disks in $\mathbb{H}^2$ in which the density of any circular disk in its respective Voronoi cell is significantly larger than $\frac{\pi}{\sqrt{12}}$. Also, we note that the constant $\frac{\pi}{\sqrt{12}}$ is best possible in the above theorem. As an additional point we mention that because the standard methods do not give a good definition of density in $\mathbb{H}^2$ (in fact all of them fail to work as was observed by Böröczky [88]) and because even today we know only a rather "fancy" way of defining density in hyperbolic space (see the work of Bowen and Radin [87]), it seems important to study finite packings in bounded containers of the hyperbolic space where there is no complication with the proper definition of density. All this leads us to the following question.

Problem 1.5.2 *Let $r > 0$ be given. Then prove or disprove that there exists a positive integer $N(r)$ with the property that the density of at least $N(r)$*

non-overlapping balls of radii r in a ball of $\mathbb{H}^3$ is always less than $\frac{\pi}{\sqrt{18}} = 0.74048\ldots$.

1.6 Upper Bounds on Sphere Packings in High Dimensions

Recall that a family of non-overlapping d-dimensional balls of radii 1 in the d-dimensional Euclidean space $\mathbb{E}^d$ is called a unit ball packing of $\mathbb{E}^d$. The density of the packing is the proportion of space covered by these unit balls. The sphere packing problem asks for the densest packing of unit balls in $\mathbb{E}^d$. Indubitably, of all problems concerning packing it was the sphere packing problem which attracted the most attention in the past decade. It has its roots in geometry, number theory, and information theory and it is part of Hilbert's 18th problem. The reader is referred to [108] (especially the third edition, which has about 800 references covering 1988-1998) for further information, definitions, and references. In what follows we report on a few selected recent developments.

The lower bound for the volume of Voronoi cells in congruent sphere packings due to C. A. Rogers [218] and Baranovskii [20] (mentioned earlier) combined with Daniel's asymptotic formula [219] yields the following corollary.

Theorem 1.6.1 *The (upper) density of any unit ball packing in $\mathbb{E}^d$ is at most*

$$\sigma_d = \frac{d}{e} 2^{-(0.5+o(1))d} \quad \text{(as } d \to \infty).$$

Then 20 years later, in 1978 Kabatiansky and Levenshtein [173] improved this bound in the exponential order of magnitude as follows. They proved the following theorem.

Theorem 1.6.2 *The (upper) density of any unit ball packing in $\mathbb{E}^d$ is at most*

$$2^{-(0.599+o(1))d} \quad \text{(as } d \to \infty).$$

In fact, Rogers' bound is better than the Kabatiansky–Levenshtein bound for $4 \le d \le 42$ and above that the Kabatiansky–Levenshtein bound takes over ([108], p. 20).

There has been some very important recent progress concerning the existence of economical packings. On the one hand, improving earlier results, Ball [13] proved the following statement through a very elegant completely new variational argument. (See also [152] for a similar result of W. Schmidt on centrally symmetric convex bodies.)

Theorem 1.6.3 *For each d, there is a lattice packing of unit balls in $\mathbb{E}^d$ with density at least*

$$\frac{d-1}{2^{d-1}} \zeta(d),$$

where $\zeta(d) = \sum_{k=1}^{\infty} \frac{1}{k^d}$ is the Riemann zeta function.

On the other hand, for some small values of d, there are explicit (lattice) packings which give (considerably) higher densities than the bound just stated. The reader is referred to [108] and [209] for a comprehensive view of results of this type.

Further improvements on the upper bounds $\widehat{\sigma}_d < \sigma_d$ for the dimensions from 4 to 36 have been obtained very recently by Cohn and Elkies [103]. They developed an analogue for sphere packing of the linear programming bounds for error-correcting codes, and used it to prove new upper bounds for the density of sphere packings, which are better than the author's upper bounds $\widehat{\sigma}_d$ for the dimensions 4 through 36. Their method together with the best-known sphere packings yields the following nearly optimal estimates in dimensions 8 and 24.

Theorem 1.6.4 *The density of the densest unit ball packing in* $\mathbb{E}^8$ *(resp.,* $\mathbb{E}^{24}$*) is at least* $0.2536\ldots$ *(resp.,* $0.00192\ldots$*) and is at most* $0.2537\ldots$ *(resp.,* $0.00196\ldots$*).*

Cohn and Elkies [103] conjecture that their approach can be used to solve the sphere packing problem in $\mathbb{E}^8$ (resp., $\mathbb{E}^{24}$).

Conjecture 1.6.5 *The* E_8 *root lattice (resp., the Leech lattice) that produces the corresponding lower bound in the previous theorem in fact, represents the largest possible density for unit sphere packings in* $\mathbb{E}^8$ *(resp.,* $\mathbb{E}^{24}$*).*

If linear programming bounds can indeed be used to prove optimality of these lattices, it would not come as a complete surprise because, for example, the kissing number problem in these dimensions was solved similarly.

Last but not least we mention the following more recent and related result of Cohn and Kumar [104] according to which the Leech lattice is the densest lattice packing in $\mathbb{E}^{24}$. (The densest lattices have been known up to dimension 8.)

Theorem 1.6.6 *The Leech lattice is the unique densest lattice in* $\mathbb{E}^{24}$*, up to scaling and isometries of* $\mathbb{E}^{24}$*.*

We close this section with a short summary on the recent progress of L. Fejes Tóth's [138] "sausage conjecture" that is one of the main problems of the theory of finite sphere packings. According to this conjecture if in $\mathbb{E}^d$, $d \geq 5$ we take $n \geq 1$ non-overlapping unit balls, then the volume of their convex hull is at least as large as the volume of the convex hull of the "sausage arrangement" of n non-overlapping unit balls under which we mean an arrangement whose centers lie on a line of $\mathbb{E}^d$ such that the unit balls of any two consecutive centers touch each other. By optimizing the methods developed by Betke, Henk, and Wills [28], [29], finally Betke and Henk [27] succeeded in proving the sausage conjecture of L. Fejes Tóth in any dimension of at least 42. Thus, we have the following natural-looking but, extremely not trivial theorem.

Theorem 1.6.7 *The sausage conjecture holds in* $\mathbb{E}^d$ *for all* $d \geq 42$*.*

It remains a highly interesting challenge to prove or disprove the sausage conjecture of L. Fejes Tóth for the dimensions between 5 and 41.

Conjecture 1.6.8 *Let* $5 \leq d \leq 41$ *be given. Then the volume of the convex hull of* $n \geq 1$ *non-overlapping unit balls in* $\mathbb{E}^d$ *is at least as large as the volume of the convex hull of the "sausage arrangement" of* n *non-overlapping unit balls which is an arrangement whose centers lie on a line of* $\mathbb{E}^d$ *such that the unit balls of any two consecutive centers touch each other.*

1.7 Uniform Stability of Sphere Packings

The notion of solidity, introduced by L. Fejes Tóth [136] to overcome difficulties of the proper definition of density in the hyperbolic plane, has been proved very useful and stimulating. Roughly speaking, a family of convex sets generating a packing is said to be *solid* if no proper rearrangement of any finite subset of the packing elements can provide a packing. More concretely, a circle packing in the plane of constant curvature is called solid if no finite subset of the circles can be rearranged such that the rearranged circles together with the rest of the circles form a packing not congruent to the original. An (easy) example for solid circle packings is the family of incircles of a regular tiling $\{p, 3\}$ for any $p \geq 3$. In fact, a closer look at this example led L. Fejes Tóth [140] to the following simple sounding but difficult problem: he conjectured that the incircles of a regular tiling $\{p, 3\}$ form a strongly solid packing for any $p \geq 5$; that is, by removing any circle from the packing the remaining circles still form a solid packing. This conjecture has been verified for $p = 5$ by Böröczky [90] and Danzer [117] and for $p \geq 8$ by A. Bezdek [30]. Thus, we have the following theorem.

Theorem 1.7.1 *The incircles of a regular tiling* $\{p, 3\}$ *form a strongly solid packing for* $p = 5$ *and for any* $p \geq 8$.

The outstanding open question left is the following.

Conjecture 1.7.2 *The incircles of a regular tiling* $\{p, 3\}$ *form a strongly solid packing for* $p = 6$ *as well as for* $p = 7$.

In connection with solidity and finite stability (of circle packings) the notion of uniform stability (of sphere packings) has been introduced by the author, A. Bezdek, and Connelly [34]. According to this a sphere packing (in the space of constant curvature) is said to be *uniformly stable* if there exists an $\epsilon > 0$ such that no finite subset of the balls of the packing can be rearranged such that each ball is moved by a distance less than ϵ and the rearranged balls together with the rest of the balls form a packing not congruent to the original one. Now, suppose that $\mathcal{P}$ is a packing of (not necessarily) congruent balls in $\mathbb{E}^d$. Let $G_\mathcal{P}$ be the contact graph of $\mathcal{P}$, where the centers of the balls serve as the vertices of $G_\mathcal{P}$ and an edge is placed between two vertices when the corresponding two balls are tangent. The following basic principle can be used to show that many packings are uniformly stable. For the more technical definitions of "critical volume condition" and "infinitesimal rigidity" entering in the theorem below we refer the interested reader to the proof discussed in the proper section of this book.

Theorem 1.7.3 *Suppose that* $\mathbb{E}^d, d \geq 2$ *can be tiled face-to-face by congruent copies of finitely many convex polytopes* $\mathbf{P}_1, \mathbf{P}_2, \ldots, \mathbf{P}_m$ *such that the vertices and edges of that tiling form the vertex and edge system of the contact graph* $G_\mathcal{P}$ *of some ball packing* $\mathcal{P}$ *in* $\mathbb{E}^d$. *Assume that each* $\mathbf{P}_i$ *and the graph* $G_\mathcal{P}$ *restricted to the vertices of* $\mathbf{P}_i$ *(and regarded as a strut graph), satisfy the critical volume condition and assume that the bar framework* $\overline{G}_\mathcal{P}$ *(restricted to the vertices of* $\mathbf{P}_i$*) is infinitesimally rigid. Then the packing* $\mathcal{P}$ *is uniformly stable.*

By taking a closer look at the Delaunay tilings of a number of lattice sphere packings one can derive the following corollary (for more details see [34]).

Corollary 1.7.4 *The densest lattice sphere packings* $A_2, A_3, D_4, D_5, E_6, E_7,$ E_8 *up to dimension 8 are all uniformly stable.*

Last we mention another corollary (for details see [34]), which was also observed by Bárány and Dolbilin [23] and which supports the above-mentioned conjecture of L. Fejes Tóth.

Corollary 1.7.5 *Consider the triangular packing of circular disks of equal radii in* $\mathbb{E}^2$ *where each disk is tangent to exactly six others. Remove one disk to obtain the packing* $\mathcal{P}'$. *Then the packing* $\mathcal{P}'$ *is uniformly stable.*

2

Finite Packings by Translates of Convex Bodies

2.1 Hadwiger Numbers of Convex Bodies

Let $\mathbf{K}$ be a convex body (i.e., a compact convex set with nonempty interior) in d-dimensional Euclidean space $\mathbb{E}^d$, $d \geq 2$. Then the *Hadwiger number* $H(\mathbf{K})$ of $\mathbf{K}$ is the largest number of non-overlapping translates of $\mathbf{K}$ that can all touch $\mathbf{K}$. An elegant observation of Hadwiger [154] is the following.

Theorem 2.1.1 *For every d-dimensional convex body $\mathbf{K}$,*

$$H(\mathbf{K}) \leq 3^d - 1,$$

where equality holds if and only if $\mathbf{K}$ is an affine d-cube.

On the other hand, in another elegant paper Swinnerton–Dyer [236] proved the following lower bound for Hadwiger numbers of convex bodies.

Theorem 2.1.2 *For every d-dimensional $(d \geq 2)$convex body $\mathbf{K}$,*

$$d^2 + d \leq H(\mathbf{K}).$$

Actually, finding a better lower bound for Hadwiger numbers of d-dimensional convex bodies is a highly challenging open problem for all $d \geq 4$. (It is not hard to see that the above theorem of Swinnerton–Dyer is sharp for dimensions 2 and 3.) The best lower bound known in dimensions $d \geq 4$ is due to Talata [239], who by applying Dvoretzky's theorem on spherical sections of centrally symmetric convex bodies succeeded in showing the following inequality.

Theorem 2.1.3 *There exists an absolute constant $c > 0$ such that*

$$2^{cd} \leq H(\mathbf{K})$$

holds for every positive integer d and for every d-dimensional convex body $\mathbf{K}$.

K. Bezdek, *Classical Topics in Discrete Geometry*, CMS Books in Mathematics,
DOI 10.1007/978-1-4419-0600-7_2, © Springer Science+Business Media, LLC 2010

Now, if we look at convex bodies different from a Euclidean ball in dimensions larger than two, then our understanding of their Hadwiger numbers is very limited. Namely, we know the Hadwiger numbers of the following convex bodies different from a ball. The result for tetrahedra is due to Talata [241] and the rest was proved by Larman and Zong [187].

Theorem 2.1.4 *The Hadwiger numbers of tetrahedra, octahedra, and rhombic dodecahedra are all equal to 18.*

In order to gain some more insight on Hadwiger numbers it is natural to pose the following question.

Problem 2.1.5 *For what integers k with $12 \leq k \leq 26$ does there exist a 3-dimensional convex body with Hadwiger number k? What is the Hadwiger number of a d-dimensional simplex (resp., crosspolytope) for $d \geq 4$?*

2.2 One-Sided Hadwiger Numbers of Convex Bodies

The author and Brass [60] assigned to each convex body $\mathbf{K}$ in $\mathbb{E}^d$ a specific positive integer called the *one-sided Hadwiger number* $h(\mathbf{K})$ as follows: $h(\mathbf{K})$ is the largest number of non-overlapping translates of $\mathbf{K}$ that touch $\mathbf{K}$ and that all lie in a closed supporting halfspace of $\mathbf{K}$. In [60], using the Brunn–Minkowski inequality, the author and Brass proved the following sharp upper bound for the one-sided Hadwiger numbers of convex bodies.

Theorem 2.2.1 *If $\mathbf{K}$ is an arbitrary convex body in $\mathbb{E}^d, d \geq 2$, then*

$$h(\mathbf{K}) \leq 2 \cdot 3^{d-1} - 1.$$

Moreover, equality is attained if and only if $\mathbf{K}$ is a d-dimensional affine cube.

The following is an open problem raised in [60].

Problem 2.2.2 *Find the smallest positive integer $n(d)$ with the property that if $\mathbf{K}$ is an arbitrary convex body in $\mathbb{E}^d$, then the maximum number of non-overlapping translates of $\mathbf{K}$ that can touch $\mathbf{K}$ and can lie in an open supporting halfspace of $\mathbf{K}$ is at most $n(d)$.*

The notion of one-sided Hadwiger numbers was introduced to study the (discrete) geometry of the so-called k^+-neighbour packings, which are packings of translates of a given convex body in $\mathbb{E}^d$ with the property that each packing element is touched by at least k others from the packing, where k is a given positive integer. As this area of discrete geometry has a rather large literature we refer the interested reader to [60] for a brief survey on the relevant results. Here, we emphasize the following corollary of the previous theorem proved in [60].

Theorem 2.2.3 *If* **K** *is an arbitrary convex body in* $\mathbb{E}^d$, *then any* k^+-*neighbour packing by translates of* **K** *with* $k \geq 2 \cdot 3^{d-1}$ *must have a positive density in* $\mathbb{E}^d$. *Moreover, there is a* $(2 \cdot 3^{d-1} - 1)^+$-*neighbour packing by translates of a* d-*dimensional affine cube with density* 0 *in* $\mathbb{E}^d$.

2.3 Touching Numbers of Convex Bodies

The *touching number* $t(\mathbf{K})$ of a convex body **K** in d-dimensional Euclidean space $\mathbb{E}^d$ is the largest possible number of mutually touching translates of **K** lying in $\mathbb{E}^d$. The elegant paper [116] of Danzer and Grünbaum gives a proof of the following fundamental inequality. In fact, this inequality was phrased by Petty [213] as well as by P. Soltan [231] in another equivalent form saying that the cardinality of an equilateral set in any d-dimensional normed space is at most 2^d.

Theorem 2.3.1 *For an arbitrary convex body* **K** *of* $\mathbb{E}^d$,

$$t(\mathbf{K}) \leq 2^d$$

with equality if and only if **K** *is an affine* d-*cube.*

In connection with the above inequality the author and Pach [41] conjecture the following even stronger result.

Conjecture 2.3.2 *For any convex body* **K** *in* $\mathbb{E}^d$, $d \geq 3$ *the maximum number of pairwise tangent positively homothetic copies of* **K** *is not more than* 2^d.

Quite surprisingly this problem is still open. In [41] it was noted that $3^d - 1$ is an easy upper bound for the quantity introduced in Conjecture 2.3.2. More recently Naszódi [207] (resp., Naszódi and Lángi [208]) improved this upper bound to 2^{d+1} in the case of a general convex body (resp., to $3 \cdot 2^{d-1}$ in the case of a centrally symmetric convex body).

It is natural to ask for a non-trivial lower bound for $t(\mathbf{K})$. Brass [94], as an application of Dvoretzky's well-known theorem, gave a partial answer for the existence of such a lower bound.

Theorem 2.3.3 *For each* k *there exists a* $d(k)$ *such that for any convex body* **K** *of* $\mathbb{E}^d$ *with* $d \geq d(k)$,

$$k \leq t(\mathbf{K}).$$

It is remarkable that the natural sounding conjecture of Petty [213] stated next is still open for all $d \geq 4$.

Conjecture 2.3.4 *For each convex body* **K** *of* $\mathbb{E}^d$, $d \geq 4$,

$$d + 1 \leq t(\mathbf{K}).$$

A generalization of the concept of touching numbers was introduced by the author, Naszódi, and Visy [59] as follows. The mth *touching number* (or the mth *Petty number*) $t(m, \mathbf{K})$ of a convex body $\mathbf{K}$ of $\mathbb{E}^d$ is the largest cardinality of (possibly overlapping) translates of $\mathbf{K}$ in $\mathbb{E}^d$ such that among any m translates there are always two touching ones. Note that $t(2, \mathbf{K}) = t(\mathbf{K})$. The following theorem proved by the author, Naszódi, and Visy [59] states some upper bounds for $t(m, \mathbf{K})$.

Theorem 2.3.5 *Let $t(\mathbf{K})$ be an arbitrary convex body in $\mathbb{E}^d$. Then*

$$t(m, \mathbf{K}) \leq \min\left\{ 4^d(m - 1), \binom{2^d + m - 1}{2^d} \right\}$$

holds for all $m \geq 2$, $d \geq 2$. Also, we have the inequalities

$$t(3, \mathbf{K}) \leq 2 \cdot 3^d, \ t(m, \mathbf{K}) \leq (m - 1)[(m - 1)3^d - (m - 2)]$$

for all $m \geq 4$, $d \geq 2$. Moreover, if $\mathbf{B^d}$ (resp., $\mathbf{C^d}$) denotes a d-dimensional ball (resp., d-dimensional affine cube) of $\mathbb{E}^d$, then

$$t(2, \mathbf{B^d}) = d + 1, \ t(m, \mathbf{B^d}) \leq (m - 1)3^d, \ t(m, \mathbf{C^d}) = (m - 1)2^d$$

hold for all $m \geq 2$, $d \geq 2$.

We cannot resist raising the following question (for more details see [59]).

Problem 2.3.6 *Prove or disprove that if $\mathbf{K}$ is an arbitrary convex body in $\mathbb{E}^d$ with $d \geq 2$ and $m > 2$, then*

$$(m - 1)(d + 1) \leq t(m, \mathbf{K}) \leq (m - 1)2^d.$$

2.4 On the Number of Touching Pairs in Finite Packings

Let $\mathbf{K}$ be an arbitrary convex body in $\mathbb{E}^d$. Then the contact graph of an arbitrary finite packing by non-overlapping translates of $\mathbf{K}$ in $\mathbb{E}^d$ is the (simple) graph whose vertices correspond to the packing elements and whose two vertices are connected by an edge if and only if the corresponding two packing elements touch each other. One of the most basic questions on contact graphs is to find out the maximum number of edges that a contact graph of n non-overlapping translates of the given convex body $\mathbf{K}$ can have in $\mathbb{E}^d$. In a very recent paper [95] Brass extended the earlier mentioned result of Harborth [166] to the "unit circular disk packings" of normed planes as follows.

Theorem 2.4.1 *The maximum number of touching pairs in a packing of n translates of a convex domain $\mathbf{K}$ in $\mathbb{E}^2$ is $\lfloor 3n - \sqrt{12n - 3} \rfloor$, if $\mathbf{K}$ is not a parallelogram, and $\lfloor 4n - \sqrt{28n - 12} \rfloor$, if $\mathbf{K}$ is a parallelogram.*

The main result of this section is an upper bound for the number of touching pairs in an arbitrary finite packing of translates of a convex body, proved by the author in [57]. In order to state the theorem in question in a concise way we need a bit of notation. Let $\mathbf{K}$ be an arbitrary convex body in $\mathbb{E}^d$, $d \geq 3$. Then let $\delta(\mathbf{K})$ denote the density of a densest packing of translates of the convex body $\mathbf{K}$ in $\mathbb{E}^d$, $d \geq 3$. Moreover, let

$$\mathrm{iq}(\mathbf{K}) := \frac{(\mathrm{svol}_{d-1}(\mathrm{bd}\mathbf{K}))^d}{(\mathrm{vol}_d(\mathbf{K}))^{d-1}}$$

be the isoperimetric quotient of the convex body $\mathbf{K}$, where $\mathrm{svol}_{d-1}(\mathrm{bd}\mathbf{K})$ denotes the $(d-1)$-dimensional surface volume of the boundary $\mathrm{bd}\mathbf{K}$ of $\mathbf{K}$ and $\mathrm{vol}_d(\mathbf{K})$ denotes the d-dimensional volume of $\mathbf{K}$. Moreover, let $\mathbf{B}$ denote the closed d-dimensional ball of radius 1 centered at the origin $\mathbf{o}$ in $\mathbb{E}^d$. Finally, let $\mathbf{K_o} := \frac{1}{2}(\mathbf{K} + (-\mathbf{K}))$ be the normalized (centrally symmetric) difference body assigned to $\mathbf{K}$ with $H(\mathbf{K_o})$ (resp., $h(\mathbf{K_o})$) standing for the Hadwiger number (resp., one-sided Hadwiger number) of $\mathbf{K_o}$.

Theorem 2.4.2 *The number of touching pairs in an arbitrary packing of $n > 1$ translates of the convex body $\mathbf{K}$ in $\mathbb{E}^d$, $d \geq 3$ is at most*

$$\frac{H(\mathbf{K_o})}{2} n - \frac{1}{2^d \delta(\mathbf{K_o})^{\frac{d-1}{d}}} \sqrt[d]{\frac{\mathrm{iq}(\mathbf{B})}{\mathrm{iq}(\mathbf{K_o})}} n^{\frac{d-1}{d}} - (H(\mathbf{K_o}) - h(\mathbf{K_o}) - 1).$$

In particular, the number of touching pairs in an arbitrary packing of $n > 1$ translates of a convex body in $\mathbb{E}^d$, $d \geq 3$ is at most

$$\frac{3^d - 1}{2} n - \frac{\sqrt[d]{\omega_d}}{2^{d+1}} n^{\frac{d-1}{d}},$$

where $\omega_d = \frac{\pi^{\frac{d}{2}}}{\Gamma(\frac{d}{2}+1)}$ is the volume of a d-dimensional ball of radius 1 in $\mathbb{E}^d$.

In the proof of Theorem 2.4.2 published by the author [57] the following statement plays an important role that might be of independent interest and so we quote it as follows. For the sake of completeness we wish to point out that Theorem 2.4.3 and Corollary 2.4.4, are actual strengthenings of Theorem 3.1 and Corollary 3.1 of [28] mainly because, in our case the containers of the packings in question are highly non-convex.

Theorem 2.4.3 *Let $\mathbf{K_o}$ be a convex body in $\mathbb{E}^d$, $d \geq 2$ symmetric about the origin $\mathbf{o}$ of $\mathbb{E}^d$ and let $\{\mathbf{c}_1 + \mathbf{K_o}, \mathbf{c}_2 + \mathbf{K_o}, \ldots, \mathbf{c}_n + \mathbf{K_o}\}$ be an arbitrary packing of $n > 1$ translates of $\mathbf{K_o}$ in $\mathbb{E}^d$. Then*

$$\frac{n\mathrm{vol}_d(\mathbf{K_o})}{\mathrm{vol}_d(\bigcup_{i=1}^n \mathbf{c}_i + 2\mathbf{K_o})} \leq \delta(\mathbf{K_o}).$$

We close this section with the following immediate corollary of Theorem 2.4.3.

Corollary 2.4.4 *Let* $\mathcal{P}_n(\mathbf{K_o})$ *be the family of all possible packings of* $n > 1$ *translates of the* **o***-symmetric convex body* $\mathbf{K_o}$ *in* $\mathbb{E}^d$, $d \geq 2$. *Moreover, let*

$$\delta(\mathbf{K_o}, n) := \max\left\{ \frac{n\mathrm{vol}_d(\mathbf{K_o})}{\mathrm{vol}_d(\bigcup_{i=1}^n \mathbf{c}_i + 2\mathbf{K_o})} \mid \{\mathbf{c}_1 + \mathbf{K_o}, \ldots, \mathbf{c}_n + \mathbf{K_o}\} \in \mathcal{P}_n(\mathbf{K_o}) \right\}.$$

Then

$$\limsup_{n \to \infty} \delta(\mathbf{K_o}, n) = \delta(\mathbf{K_o}).$$

3

Coverings by Homothetic Bodies - Illumination and Related Topics

3.1 The Illumination Conjecture

Let $\mathbf{K}$ be a convex body (i.e., a compact convex set with nonempty interior) in the d-dimensional Euclidean space $\mathbb{E}^d$, $d \geq 2$. According to Hadwiger [155] an exterior point $\mathbf{p} \in \mathbb{E}^d \setminus \mathbf{K}$ of $\mathbf{K}$ illuminates the boundary point $\mathbf{q}$ of $\mathbf{K}$ if the halfline emanating from $\mathbf{p}$ passing through $\mathbf{q}$ intersects the interior of $\mathbf{K}$ (at a point not between $\mathbf{p}$ and $\mathbf{q}$). Furthermore, a family of exterior points of $\mathbf{K}$ say, $\mathbf{p}_1, \mathbf{p}_2, \ldots, \mathbf{p}_n$ illuminates $\mathbf{K}$ if each boundary point of $\mathbf{K}$ is illuminated by at least one of the point sources $\mathbf{p}_1, \mathbf{p}_2, \ldots, \mathbf{p}_n$. Finally, the smallest n for which there exist n exterior points of $\mathbf{K}$ that illuminate $\mathbf{K}$ is called the *illumination number* of $\mathbf{K}$ denoted by $I(\mathbf{K})$. In 1960, Hadwiger [155] raised the following amazingly elementary, but very fundamental question. An equivalent but somewhat different-looking concept of illumination was introduced by Boltyanski in [78]. There he proposed to use directions (i.e., unit vectors) instead of point sources for the illumination of convex bodies. Based on these circumstances we call the following conjecture the *Boltyanski–Hadwiger Illumination Conjecture*.

Conjecture 3.1.1 *The illumination number $I(\mathbf{K})$ of any convex body $\mathbf{K}$ in $\mathbb{E}^d$, $d \geq 3$ is at most 2^d and $I(\mathbf{K}) = 2^d$ if and only if $\mathbf{K}$ is an affine d-cube.*

It is quite easy to prove the Illumination Conjecture in the plane (see for example [47]). Also, it has been noticed by several people that the illumination number of any smooth convex body in $\mathbb{E}^d$ is exactly $d + 1$ ([47]). (A convex body of $\mathbb{E}^d$ is called smooth if through each of its boundary points there exists a uniquely defined supporting hyperplane of $\mathbb{E}^d$.) However, the Illumination Conjecture is widely open for convex d-polytopes as well as for non-smooth convex bodies in $\mathbb{E}^d$ for all $d \geq 3$. In fact, a proof of the Illumination Conjecture for polytopes alone would not immediately imply its correctness for convex bodies in general, mainly because of the so-called upper semicontinuity of the illumination numbers of convex bodies. More exactly, here we refer to the following statement ([84]).

K. Bezdek, *Classical Topics in Discrete Geometry*, CMS Books in Mathematics,
DOI 10.1007/978-1-4419-0600-7_3, © Springer Science+Business Media, LLC 2010

Theorem 3.1.2 *Let* **K** *be a convex body in* $\mathbb{E}^d$. *Then for any convex body* **K**′ *sufficiently close to* **K** *in the Hausdorff metric of the convex bodies in* $\mathbb{E}^d$ *the inequality* $I(\mathbf{K}') \leq I(\mathbf{K})$ *holds (often with strict inequality).*

In what follows we survey the major results known about the Illumination Conjecture. For earlier and by now less updated accounts on the status of this problem we refer the reader to the survey papers [47] and [195].

3.2 Equivalent Formulations

There are two equivalent formulations of the Illumination Conjecture that are often used in the literature (for more details see [195]). The first of these was raised by Gohberg and Markus [148]. (In fact, they came up with their problem independently of Boltyanski and Hadwiger by studying some geometric properties of normed spaces.) It is called the *Gohberg–Markus Covering Conjecture*.

Conjecture 3.2.1 *Let* **K** *be an arbitrary convex body in* $\mathbb{E}^d$, $d \geq 3$. *Then* **K** *can be covered by* 2^d *smaller positively homothetic copies and* 2^d *copies are needed only if* **K** *is an affine d-cube.*

Another equivalent formulation was found independently by P. Soltan and V. Soltan [232] (who formulated it for the centrally symmetric case only) and by the author [43] (see also [44]). In the formulation below of the *K. Bezdek – P. Soltan – V. Soltan Separation Conjecture* a face of a convex body means the intersection of the convex body with a supporting hyperplane.

Conjecture 3.2.2 *Let* **K** *be an arbitrary convex body in* $\mathbb{E}^d$, $d \geq 3$ *and* **o** *an arbitrary interior point of* **K**. *Then there exist* 2^d *hyperplanes of* $\mathbb{E}^d$ *such that each face of* **K** *can be strictly separated from* **o** *by at least one of the* 2^d *hyperplanes. Furthermore,* 2^d *hyperplanes are needed only if* **K** *is the convex hull of d linearly independent line segments which intersect at the common relative interior point* **o**.

3.3 The Illumination Conjecture in Dimension Three

The best upper bound known on the illumination numbers of convex bodies in $\mathbb{E}^3$ is due to Papadoperakis [212].

Theorem 3.3.1 *The illumination number of any convex body in* $\mathbb{E}^3$ *is at most* 16.

It is quite encouraging that the Illumination Conjecture is known to hold for some "relatively large" classes of convex bodies in $\mathbb{E}^3$ as well as in $\mathbb{E}^d$, $d \geq 4$. In what follows, first we survey the 3-dimensional results.

The author [43] succeeded in proving the following theorem.

Theorem 3.3.2 *If* **P** *is a convex polyhedron of* $\mathbb{E}^3$ *with affine symmetry (i.e., if the affine symmetry group of* **P** *consists of the identity and at least one other affinity of* $\mathbb{E}^3$*), then the illumination number of* **P** *is at most 8.*

On the other hand, the following theorem also holds. The first part of that was proved by Lassak [184] (in fact, this paper was published before the publication of Theorem 3.3.2), and the second part by Dekster [120], extending the above theorem of the author on polyhedra to convex bodies with center or plane symmetry.

Theorem 3.3.3
(i) If **K** *is a centrally symmetric convex body in* $\mathbb{E}^3$*, then* $I(\mathbf{K}) \leq 8$*.*
(ii) If **K** *is a convex body symmetric about a plane in* $\mathbb{E}^3$*, then* $I(\mathbf{K}) \leq 8$*.*

Lassak [186] and later also Weissbach [246] and the author, Lángi, Naszódi, and Papez [69] gave a proof of the following.

Theorem 3.3.4 *The illumination number of any convex body of constant width in* $\mathbb{E}^3$ *is at most 6.*

It is tempting to conjecture the following even stronger result. If true, then it would give a new proof and insight of the well-known theorem, conjectured by Borsuk long ago (see for example [1]), that any set of diameter 1 in $\mathbb{E}^3$ can be partitioned into (at most) four subsets of diameter smaller than 1.

Conjecture 3.3.5 *The illumination number of any convex body of constant width in* $\mathbb{E}^3$ *is exactly 4.*

As a last remark we need to mention the following. In [81] Boltyanski announced a solution of the Illumination Conjecture in dimension 3. Unfortunately, even today the proposed proof of this result remains incomplete. In other words, one has to regard the Illumination Conjecture as a still open problem in dimension 3.

3.4 The Illumination Conjecture in High Dimensions

It was rather a coincidence, at least from the point of view of the Illumination Conjecture, when in 1964 Erdős and Rogers [127] proved the following theorem. In order to state their theorem in a proper form we need to introduce the following notion. If we are given a covering of a space by a system of sets, the *star number* of the covering is the supremum, over sets of the system, of the cardinals of the numbers of sets of the system meeting a set of the system. On the one hand, the standard Lebesgue "brick-laying" construction provides an example, for each positive integer d, of a lattice covering of $\mathbb{E}^d$ by closed cubes with star number $2^{d+1} - 1$. On the other hand, Theorem 1 of [127] states that the star number of a lattice covering of $\mathbb{E}^d$ by translates of a centrally symmetric convex body is always at least $2^{d+1} - 1$. However, from our point of view, the main result of [127] is the one under Theorem 2 (which combined with some observations from [126] and with the inequality of Rogers and Shephard [220] on the volume of difference bodies) reads as follows.

Theorem 3.4.1 *Let* **K** *be a convex body in the d-dimensional Euclidean space* $\mathbb{E}^d, d \geq 2$. *Then there exists a covering of* $\mathbb{E}^d$ *by translates of* **K** *with star number at most*

$$\frac{\text{vol}_d(\mathbf{K} - \mathbf{K})}{\text{vol}_d(\mathbf{K})}(d \ln d + d \ln \ln d + 5d + 1) \leq \binom{2d}{d}(d \ln d + d \ln \ln d + 5d + 1).$$

Moreover, for sufficiently large d, 5d can be replaced by 4d.

The periodic and probabilistic construction on which Theorem 3.4.1 is based gives also the following.

Corollary 3.4.2 *If* **K** *is an arbitrary convex body in the d-dimensional Euclidean space* $\mathbb{E}^d$, $d \geq 2$, *then*

$$I(\mathbf{K}) \leq \frac{\text{vol}_d(\mathbf{K} - \mathbf{K})}{\text{vol}_d(\mathbf{K})}(d \ln d + d \ln \ln d + 5d) \leq \binom{2d}{d}(d \ln d + d \ln \ln d + 5d).$$

Moreover, for sufficiently large d, 5d can be replaced by 4d.

For the sake of completeness we mention also the inequality $I(\mathbf{K}) \leq (d + 1)d^{d-1} - (d - 1)(d - 2)^{d-1}$ due to Lassak [185], which is valid for an arbitrary convex body **K** in $\mathbb{E}^d$, $d \geq 2$. (Actually, Lassak's estimate is (somewhat) better than the estimate of Corollary 3.4.2 for some small values of d.)

Note that, from the point of view of the Illumination Conjecture, the estimate of Corollary 3.4.2 is nearly best possible for centrally symmetric convex bodies, because in that case $\frac{\text{vol}_d(\mathbf{K}-\mathbf{K})}{\text{vol}_d(\mathbf{K})} = 2^d$. However, most convex bodies are far from being symmetric and so, in general, one may wonder whether the Illumination Conjecture is true at all, in particular, in high dimensions. Thus, it was important progress when Schramm [226] managed to prove the Illumination Conjecture for all convex bodies of constant width in all dimensions at least 16. In fact, he has proved the following inequality.

Theorem 3.4.3 *If* **W** *is an arbitrary convex body of constant width in* $\mathbb{E}^d, d \geq 3$, *then*

$$I(\mathbf{W}) < 5d\sqrt{d}(4 + \ln d)\left(\frac{3}{2}\right)^{\frac{d}{2}}.$$

Very recently the author has extended the estimate of Theorem 3.4.3 to a family of convex bodies much larger than the family of convex bodies of constant width also including the family of "fat" *ball-polyhedra*. For the more exact details see Theorem 6.8.3 and also the discussion there on the illumination of ball-polyhedra, which are convex bodies that with scaling form an everywhere dense subset of the space of convex bodies.

Recall that a convex polytope is called a belt polytope if to each side of any of its 2-faces there exists a parallel (opposite) side on the same 2-face. This class of polytopes is wider than the class of zonotopes, moreover, it is easy to see that any convex body of $\mathbb{E}^d$ can be represented as a limit of a covergent sequence of belt polytopes with respect to the Hausdorff metric in $\mathbb{E}^d$. The following theorem on belt polytopes was proved by Martini in [194]. The result that it extends to the class of convex bodies called belt bodies (including zonoids) is due to Boltyanski [80]. (See also [83] for a somewhat sharper result on the illumination numbers of belt bodies.)

Theorem 3.4.4 *Let* **P** *be an arbitrary d-dimensional belt polytope (resp., belt body) different from a parallelotope in $\mathbb{E}^d$, $d \geq 2$. Then*

$$I(\mathbf{P}) \leq 3 \cdot 2^{d-2}.$$

Now, let **K** be an arbitrary convex body in $\mathbb{E}^d$ and let $T(\mathbf{K})$ be the family of all translates of **K** in $\mathbb{E}^d$. The Helly dimension $\mathrm{him}(\mathbf{K})$ of **K** is the smallest integer h such that for any finite family $\mathcal{F} \subset T(\mathbf{K})$ with $\mathrm{card}\mathcal{F} > h + 1$ the following assertion holds: if every $h + 1$ members of $\mathcal{F}$ have a point in common, then all the members of $\mathcal{F}$ have a point in common. As is well known $1 \leq \mathrm{him}(\mathbf{K}) \leq d$. Using this notion Boltyanski [82] gave a proof of the following theorem.

Theorem 3.4.5 *Let* **K** *be a convex body with* $\mathrm{him}(\mathbf{K}) = 2$ *in $\mathbb{E}^d$, $d \geq 3$. Then*

$$I(\mathbf{K}) \leq 2^d - 2^{d-2}.$$

In fact, in [82] Boltyanski conjectures the following more general inequality.

Conjecture 3.4.6 *Let* **K** *be a convex body with* $\mathrm{him}(\mathbf{K}) = h > 2$ *in $\mathbb{E}^d$, $d \geq 3$. Then*

$$I(\mathbf{K}) \leq 2^d - 2^{d-h}.$$

The author and Bisztriczky gave a proof of the Illumination Conjecture for the class of dual cyclic polytopes in [53]. Their upper bound for the illumination numbers of dual cyclic polytopes has been improved by Talata in [240]. So, we have the following statement.

Theorem 3.4.7 *The illumination number of any d-dimensional dual cyclic polytope is at most $\frac{(d+1)^2}{2}$ for all $d \geq 2$.*

In connection with the results of this section quite a number of questions remain open including the following ones.

Problem 3.4.8

(i) What are the illumination numbers of cyclic polytopes?
(ii) Can one give a proof of the Separation Conjecture for zonotopes (resp., belt polytopes)?
(iii) Is there a way to prove the Separation Conjecture for 0/1-polytopes?

3.5 On the X-Ray Number of Convex Bodies

In 1972, the X-ray number of convex bodies was introduced by P. Soltan as follows (see also [195]). Let $\mathbf{K}$ be a convex body of $\mathbb{E}^d$, $d \geq 2$. Let $L \subset \mathbb{E}^d$ be a line through the origin of $\mathbb{E}^d$. We say that the point $\mathbf{p} \in \mathbf{K}$ is *X-rayed along L* if the line parallel to L passing through $\mathbf{p}$ intersects the interior of $\mathbf{K}$. The *X-ray number* $X(\mathbf{K})$ of $\mathbf{K}$ is the smallest number of lines such that every point of $\mathbf{K}$ is X-rayed along at least one of these lines. Obviously, $X(\mathbf{K}) \geq d$. Moreover, it is easy to see that this bound is attained by any smooth convex body. On the other hand, if $\mathbf{C}_d$ is a d-dimensional (affine) cube and F is one of its $(d-2)$-dimensional faces, then the X-ray number of the convex hull of the set of vertices of $\mathbf{C}_d \setminus F$ is $3 \cdot 2^{d-2}$.

In 1994, the author and Zamfirescu [52] published the following conjecture.

Conjecture 3.5.1 *The X-ray number of any convex body in $\mathbb{E}^d$ is at most $3 \cdot 2^{d-2}$.*

This conjecture, which we call the X-ray Conjecture, is proved only in the plane and it is open in high dimensions. A related and much better studied problem is the above-mentioned Illumination Conjecture. Here we note that the inequalities $X(\mathbf{K}) \leq I(\mathbf{K}) \leq 2X(\mathbf{K})$ hold for any convex body $\mathbf{K} \subset \mathbb{E}^d$. Putting it differently, any proper progress on the X-ray Conjecture would imply progress on the Illumination Conjecture and vice versa. Finally we note that a natural way to prove the X-ray Conjecture would be to show that any convex body $\mathbf{K} \subset \mathbb{E}^d$ can be illuminated by $3 \cdot 2^{d-2}$ pairs of pairwise opposite directions. The main results of [72] can be summarized as follows. In order to state it properly we need to recall two basic notions. Let $\mathbf{K}$ be a convex body in $\mathbb{E}^d$ and let F be a face of $\mathbf{K}$. The *Gauss image* $\nu(F)$ of the face F is the set of all points (i.e., unit vectors) $\mathbf{u}$ of the $(d-1)$-dimensional unit sphere $\mathbb{S}^{d-1} \subset \mathbb{E}^d$ centered at the origin $\mathbf{o}$ of $\mathbb{E}^d$ for which the supporting hyperplane of $\mathbf{K}$ with outer normal vector $\mathbf{u}$ contains F. It is easy to see that the Gauss images of distinct faces of $\mathbf{K}$ have disjoint relative interiors in $\mathbb{S}^{d-1}$ and $\nu(F)$ is compact and spherically convex for any face F. Let $C \subset \mathbb{S}^{d-1}$ be a set of finitely many points. Then the *covering radius* of C is the smallest positive real number r with the property that the family of spherical balls of radii r centered at the points of C covers $\mathbb{S}^{d-1}$.

Theorem 3.5.2 *Let $\mathbf{K} \subset \mathbb{E}^d$, $d \geq 3$ be a convex body and let r be a positive real number with the property that the Gauss image $\nu(F)$ of any face F of $\mathbf{K}$ can be covered by a spherical ball of radius r in $\mathbb{S}^{d-1}$. Moreover, assume* *that there exist $2m$ pairwise antipodal points of $\mathbb{S}^{d-1}$ with covering radius R satisfying the inequality $r + R \leq \frac{\pi}{2}$. Then $X(\mathbf{K}) \leq m$. In particular, if there are $2m$ pairwise antipodal points on $\mathbb{S}^{d-1}$ with covering radius R satisfying the inequality $R \leq \pi/2 - r_{d-1}$, where $r_{d-1} = \arccos \sqrt{\frac{d+1}{2d}}$ is the circumradius of a regular $(d-1)$-dimensional spherical simplex of edge length $\pi/3$, then $X(\mathbf{W}) \leq m$ holds for any convex body of constant width $\mathbf{W}$ in $\mathbb{E}^d$.*

Theorem 3.5.3 *If $\mathbf{W}$ is an arbitrary convex body of constant width in $\mathbb{E}^3$, then $X(\mathbf{W}) = 3$. If $\mathbf{W}$ is any convex body of constant width in $\mathbb{E}^4$, then $4 \leq X(\mathbf{W}) \leq 6$. Moreover, if $\mathbf{W}$ is a convex body of constant width in $\mathbb{E}^d$ with $d = 5, 6$, then $d \leq X(\mathbf{W}) \leq 2^{d-1}$.*

Corollary 3.5.4 *If $\mathbf{W}$ is an arbitrary convex body of constant width in $\mathbb{E}^3$, then $4 \leq I(\mathbf{W}) \leq 6$. If $\mathbf{W}$ is any convex body of constant width in $\mathbb{E}^4$, then $5 \leq I(\mathbf{W}) \leq 12$. Moreover, if $\mathbf{W}$ is a convex body of constant width in $\mathbb{E}^d$ with $d = 5, 6$, then $d + 1 \leq I(\mathbf{W}) \leq 2^d$.*

It would be interesting to extend the method described in the paper [72] for the next couple of dimensions (more exactly, for the dimensions $7 \leq d \leq 14$) in particular, because in these dimensions neither the X-ray Conjecture nor the Illumination Conjecture is known to hold for convex bodies of constant width.

3.6 The Successive Illumination Numbers of Convex Bodies

Let $\mathbf{K}$ be a convex body in $\mathbb{E}^d$, $d \geq 2$. The following definitions were introduced by the author in [50] (see also [43] that introduced the concept of the first definition below).

Let $L \subset \mathbb{E}^d \setminus \mathbf{K}$ be an affine subspace of dimension l, $0 \leq l \leq d - 1$. Then L illuminates the boundary point $\mathbf{q}$ of $\mathbf{K}$ if there exists a point $\mathbf{p}$ of L that illuminates $\mathbf{q}$ on the boundary of $\mathbf{K}$. Moreover, we say that the affine subspaces $L_1, L_2, \ldots, L_n$ of dimension l with $L_i \subset \mathbb{E}^d \setminus \mathbf{K}, 1 \leq i \leq n$ illuminate $\mathbf{K}$ if every boundary point of $\mathbf{K}$ is illuminated by at least one of the affine subspaces $L_1, L_2, \ldots, L_n$. Finally, let $I_l(\mathbf{K})$ be the smallest positive integer n for which there exist n affine subspaces of dimension l say, $L_1, L_2, \ldots, L_n$ such that $L_i \subset \mathbb{E}^d \setminus \mathbf{K}$ for all $1 \leq i \leq n$ and $L_1, L_2, \ldots, L_n$ illuminate $\mathbf{K}$. $I_l(\mathbf{K})$ is called the l-dimensional illumination number of $\mathbf{K}$ and the sequence $I_0(\mathbf{K}), I_1(\mathbf{K}), \ldots, I_{d-2}(\mathbf{K}), I_{d-1}(\mathbf{K})$ is called the *successive illumination numbers* of $\mathbf{K}$. Obviously, $I_0(\mathbf{K}) \geq I_1(\mathbf{K}) \geq \cdots \geq I_{d-2}(\mathbf{K}) \geq I_{d-1}(\mathbf{K}) = 2$.

Let $\mathbb{S}^{d-1}$ be the unit sphere centered at the origin of $\mathbb{E}^d$. Let $HS^l \subset \mathbb{S}^{d-1}$ be an l-dimensional open great-hemisphere of $\mathbb{S}^{d-1}$, where $0 \leq l \leq d - 1$. Then HS^l illuminates the boundary point $\mathbf{q}$ of $\mathbf{K}$ if there exists a unit vector $\mathbf{v} \in HS^l$ that illuminates $\mathbf{q}$, in other words, for which it is true that the halfline emanating from $\mathbf{q}$ and having direction vector $\mathbf{v}$ intersects the interior of $\mathbf{K}$. Moreover, we say that the l-dimensional open great-hemispheres $HS_1^l, HS_2^l, \ldots, HS_n^l$ of $\mathbb{S}^{d-1}$ illuminate $\mathbf{K}$ if each boundary point of $\mathbf{K}$ is illuminated by at least one of the open great-hemispheres $HS_1^l, HS_2^l, \ldots, HS_n^l$. Finally, let $I_l'(\mathbf{K})$ be the smallest number of l-dimensional open great-hemispheres of $\mathbb{S}^{d-1}$ that illuminate $\mathbf{K}$. Obviously, $I_0'(\mathbf{K}) \geq I_1'(\mathbf{K}) \geq \cdots \geq I_{d-2}'(\mathbf{K}) \geq I_{d-1}'(\mathbf{K}) = 2$.

Let $L \subset \mathbb{E}^d$ be a linear subspace of dimension l, $0 \leq l \leq d-1$ in $\mathbb{E}^d$. The lth order circumscribed cylinder of $\mathbf{K}$ generated by L is the union of translates of L that have a nonempty intersection with $\mathbf{K}$. Then let $C_l(\mathbf{K})$ be the smallest number of translates of the interiors of some lth order circumscribed cylinders of $\mathbf{K}$ the union of which contains $\mathbf{K}$. Obviously, $C_0(\mathbf{K}) \geq C_1(\mathbf{K}) \geq \cdots \geq C_{d-2}(\mathbf{K}) \geq C_{d-1}(\mathbf{K}) = 2$.

The following theorem, which was proved in [50], collects the basic information known about the quantities just introduced. (The inequality (ii) was in fact, first proved in [45] and proved again in a different way in [48].)

Theorem 3.6.1 *Let $\mathbf{K}$ be an arbitrary convex body of $\mathbb{E}^d$. Then*
(i) $I_l(\mathbf{K}) = I_l'(\mathbf{K}) = C_l(\mathbf{K})$ for all $0 \leq l \leq d-1$;
(ii) $\lceil \frac{d+1}{l+1} \rceil \leq I_l(\mathbf{K})$ for all $0 \leq l \leq d-1$ with equality for any smooth $\mathbf{K}$;
(iii) $I_{d-2}(\mathbf{K}) = 2$ for all $d \geq 3$.

The *Generalized Illumination Conjecture* was phrased by the author in [50] as follows.

Conjecture 3.6.2 *Let $\mathbf{K}$ be an arbitrary convex body and $\mathbf{C}$ be a d-dimensional affine cube in $\mathbb{E}^d$. Then*

$$I_l(\mathbf{K}) \leq I_l(\mathbf{C})$$

holds for all $0 \leq l \leq d-1$.

The above conjecture was proved for zonotopes and zonoids in [50]. The results of parts (i) and (ii) of the next theorem are taken from [50], where they were proved for zonotopes (resp., zonoids). However, in the light of the more recent works in [80] and [83] these results extend to the class of belt polytopes (resp., belt bodies) in a rather straightforward way so we present them in that form. The lower bound of part (iii) was proved in [50] and the upper bound of part (iii) is the major result of [179]. Finally, part (iv) was proved in [49].

Theorem 3.6.3 *Let $\mathbf{K}'$ be a belt polytope (resp., belt body) and $\mathbf{C}$ be a d-dimensional affine cube in $\mathbb{E}^d$. Then*
(i) $I_l(\mathbf{K}') \leq I_l(\mathbf{C})$ holds for all $0 \leq l \leq d-1$;
(ii) $I_{\lfloor \frac{d}{2} \rfloor}(\mathbf{K}') = \cdots = I_{d-1}(\mathbf{K}') = 2$;
(iii) $\frac{2^d}{\sum_{i=0}^{l} \binom{d}{i}} \leq I_l(\mathbf{C}) \leq K(d,l)$, where $K(d,l)$ denotes the minimum cardinality of binary codes of length d with covering radius l, $0 \leq l \leq d-1$;

(iv) $I_1(\mathbf{C}) = \frac{2^d}{d+1}$ provided that $d+1 = 2^m$.

We close this section with a conjecture of Kiss ([179]) on the illumination numbers of affine cubes and call the attention of the reader to the problem of proving the Generalized Illumination Conjecture in a stronger form for convex bodies of constant width in $\mathbb{E}^4$.

Conjecture 3.6.4 Let $\mathbf{C}$ be a d-dimensional affine cube in $\mathbb{E}^d$. Then $I_l(\mathbf{C}) = K(d,l)$, where $K(d,l)$ denotes the minimum cardinality of binary codes of length d with covering radius l, $0 \leq l \leq d-1$.

Conjecture 3.6.5 Let $\mathbf{K}''$ be a convex body of constant width in $\mathbb{E}^4$. Then $I_0(\mathbf{K}'') \leq 8$ (and so, $I_1(\mathbf{K}'') \leq 4$).

3.7 The Illumination and Covering Parameters of Convex Bodies

Let $\mathbf{K_o}$ be a convex body in $\mathbb{E}^d$, $d \geq 2$ symmetric about the origin $\mathbf{o}$ of $\mathbb{E}^d$. Then $\mathbf{K_o}$ defines the norm

$$\|\mathbf{x}\|_{\mathbf{K_o}} = \inf\{\lambda \mid \mathbf{x} \in \lambda \mathbf{K_o}\}$$

of any $\mathbf{x} \in \mathbb{E}^d$ (with respect to $\mathbf{K_o}$).

The *illumination parameter* ill($\mathbf{K_o}$) of $\mathbf{K_o}$ was introduced by the author in [46] as follows.

$$\mathrm{ill}(\mathbf{K_o}) = \inf\left\{\sum_i \|\mathbf{p_i}\|_{\mathbf{K_o}} \mid \{\mathbf{p_i}\} \text{ illuminates } \mathbf{K_o}\right\}.$$

Clearly this ensures that far-away light sources are penalised. The following theorem was proved in [46] (see also [67]). In the same paper the problem of finding the higher-dimensional analogue of that claim was raised as well.

Theorem 3.7.1 If $\mathbf{K_o}$ is an $\mathbf{o}$-symmetric convex domain of $\mathbb{E}^2$, then

$$\mathrm{ill}(\mathbf{K_o}) \leq 6$$

with equality for any affine regular convex hexagon.

The illumination parameters of the cube, octahedron, dodecahedron, and icosahedron (i.e., of the centrally symmetric Platonic solids) have been computed in the papers [67] and [180]. Thus, we have the following theorem.

Theorem 3.7.2 The illumination parameters of the (affine) cube, octahedron, dodecahedron, and icosahedron in $\mathbb{E}^3$ are equal to $8, 6, 4\sqrt{5} + 2$, and 12.

Kiss and de Wet [180] conjecture the following.

Conjecture 3.7.3 *The illumination parameter of any* **o**-*symmetric convex body in* $\mathbb{E}^3$ *is at most 12.*

Motivated by the notion of the illumination parameter Swanepoel [235] introduced the *covering parameter* $\text{cov}(\mathbf{K_o})$ of $\mathbf{K_o}$ in the following way.

$$\text{cov}(\mathbf{K_o}) = \inf\{\sum_i (1 - \lambda_i)^{-1} \mid \mathbf{K_o} \subset \cup_i(\lambda_i\mathbf{K_o} + \mathbf{t_i}), 0 < \lambda_i < 1, \mathbf{t_i} \in \mathbb{E}^d\}.$$

In this way homothets almost as large as $\mathbf{K_o}$ are penalised. Swanepoel [235] proved the following fundamental inequalities.

Theorem 3.7.4 *For any* **o**-*symmetric convex body* $\mathbf{K_o}$ *in* $\mathbb{E}^d$, $d \geq 2$ *we have that*
(i) $\text{ill}(\mathbf{K_o}) \leq 2\text{cov}(\mathbf{K_o}) \leq O(2^d d^2 \ln d)$;
(ii) $v(\mathbf{K_o}) \leq \text{ill}(\mathbf{K_o})$, *where* $v(\mathbf{K_o})$ *is the maximum possible degree of a vertex in a* $\mathbf{K_o}$-*Steiner minimal tree.*

Based on the above theorems, it is natural to study the following question that was proposed by Swanepoel [235]). One can regard this problem as the quantitative analogue of the Illumination Conjecture.

Problem 3.7.5 *Prove or disprove that the inequality* $\text{ill}(\mathbf{K_o}) \leq O(2^d)$ *holds for all* **o**-*symmetric convex bodies* $\mathbf{K_o}$ *of* $\mathbb{E}^d$.

3.8 On the Vertex Index of Convex Bodies

The following concept related to the illumination parameters of convex bodies was introduced by the author and Litvak in [70]. Let $\mathbf{K_o}$ be a convex body in $\mathbb{E}^d$, $d \geq 2$ symmetric about the origin **o** of $\mathbb{E}^d$. Now, place $\mathbf{K_o}$ in a convex polytope, say $\mathbf{P}$, with vertices $\mathbf{p}_1, \mathbf{p}_2, \ldots, \mathbf{p}_n$, where $n \geq d + 1$. Then it is natural to measure the closeness of the vertex set of $\mathbf{P}$ to the origin **o** by computing $\sum_{1 \leq i \leq n} \|\mathbf{p}_i\|_{\mathbf{K_o}}$, where $\|\mathbf{x}\|_{\mathbf{K_o}} = \inf\{\lambda > 0 \mid \mathbf{x} \in \lambda\mathbf{K_o}\}$ denotes the norm of $\mathbf{x} \in \mathbb{E}^d$ with respect to $\mathbf{K_o}$. Finally, look for the convex polytope that contains $\mathbf{K_o}$ and whose vertex set has the smallest possible closeness to **o** and introduce the *vertex index*, $\text{vein}(\mathbf{K_o})$, of $\mathbf{K_o}$ as follows,

$$\text{vein}(\mathbf{K_o}) = \inf\left\{\sum_i \|\mathbf{p}_i\|_{\mathbf{K_o}} \mid \mathbf{K_o} \subset \text{conv}\{\mathbf{p}_i\}\right\}.$$

Note that $\text{vein}(\mathbf{K_o})$ is an affine invariant quantity assigned to $\mathbf{K_o}$; that is, if $A : \mathbb{E}^d \to \mathbb{E}^d$ is an (invertible) linear map, then $\text{vein}(\mathbf{K_o}) = \text{vein}(A(\mathbf{K_o}))$. Moreover, it is also clear that

$$\text{vein}(\mathbf{K_o}) \le \text{ill}(\mathbf{K_o})$$

holds for any $\mathbf{o}$-symmetric convex body $\mathbf{K_o}$ in $\mathbb{E}^d$ with equality for smooth convex bodies.

In what follows we summarize the major results of [70]. Also, we note that estimates on Banach–Mazur distances between convex bodies as well as on (outer) volume ratios of convex bodies, approximation by convex polytopes and spherical isoperimetric inequalities turn out to play a central role in the proofs of the following two theorems of [70].

Theorem 3.8.1 *For every $d \ge 2$ one has*

$$\frac{d^{3/2}}{\sqrt{2\pi e}} \le \text{vein}(\mathbf{B}_2^d) \le 2d^{3/2},$$

where $\mathbf{B}_2^d$ denotes the unit ball of $\mathbb{E}^d$. Moreover, if $d = 2$, 3 then $\text{vein}(\mathbf{B}_2^d) = 2d^{3/2}$.

The author and Litvak [70] conjecture the following stonger result.

Conjecture 3.8.2 *For every $d \ge 2$ one has*

$$\text{vein}(\mathbf{B}_2^d) = 2d^{3/2}.$$

The result of [70] mentioned next gives the "big picture" presently known on vertex indices of convex bodies.

Theorem 3.8.3 *There is an absolute constant $C > 0$ such that for every $d \ge 2$ and every $\mathbf{o}$-symmetric convex body $\mathbf{K_o}$ in $\mathbb{E}^d$ one has*

$$\frac{d^{3/2}}{\sqrt{2\pi e}\ \text{ovr}(\mathbf{K_o})} \le \text{vein}(\mathbf{K_o}) \le Cd^{3/2}\ln(2d),$$

where $\text{ovr}(\mathbf{K_o}) = \inf\{\text{vol}_d(\mathcal{E})/\text{vol}_d(\mathbf{K_o})\}^{1/d}$ is the outer volume ratio of $\mathbf{K_o}$ with the infimum taken over all ellipsoids $\mathcal{E}$ containing $\mathbf{K_o}$ and with $\text{vol}_d(\dots)$ denoting the volume.

It has been noted in [70] that the above-mentioned conjecture on $\text{vein}(\mathbf{B}_2^d)$ implies the inequality $2d \le \text{vein}(\mathbf{K_o})$ for any $\mathbf{o}$-symmetric convex body $\mathbf{K_o}$ in $\mathbb{E}^d$ (with equality for d-dimensional crosspolytopes). This inequality has just been proved in [147].

Theorem 3.8.4 *Let $\mathbf{K_o}$ be an $\mathbf{o}$-symmetric convex body in $\mathbb{E}^d$. Then*

$$2d \le \text{vein}(\mathbf{K_o}).$$

4

Coverings by Planks and Cylinders

4.1 Plank Theorems

As usual, a *convex body* of the Euclidean space $\mathbb{E}^d$ is a compact convex set with non-empty interior. Let $\mathbf{C} \subset \mathbb{E}^d$ be a convex body, and let $H \subset \mathbb{E}^d$ be a hyperplane. Then the distance $w(\mathbf{C}, H)$ between the two supporting hyperplanes of $\mathbf{C}$ parallel to H is called the *width of* $\mathbf{C}$ *parallel to H*. Moreover, the smallest width of $\mathbf{C}$ parallel to hyperplanes of $\mathbb{E}^d$ is called the *minimal width* of $\mathbf{C}$ and is denoted by $w(\mathbf{C})$.

Recall that in the 1930's, Tarski posed what came to be known as the plank problem. A *plank* $\mathbf{P}$ in $\mathbb{E}^d$ is the (closed) set of points between two distinct parallel hyperplanes. The *width* $w(\mathbf{P})$ of $\mathbf{P}$ is simply the distance between the two boundary hyperplanes of $\mathbf{P}$. Tarski conjectured that if a convex body of minimal width w is covered by a collection of planks in $\mathbb{E}^d$, then the sum of the widths of these planks is at least w. This conjecture was proved by Bang in his memorable paper [18]. (In fact, the proof presented in that paper is a simplification and generalization of the proof published by Bang somewhat earlier in [17].) Thus, we call the following statement Bang's plank theorem.

Theorem 4.1.1 *If the convex body* $\mathbf{C}$ *is covered by the planks* $\mathbf{P}_1, \mathbf{P}_2, \dots, \mathbf{P}_n$ *in* $\mathbb{E}^d, d \geq 2$ *(i.e.,* $\mathbf{C} \subset \mathbf{P}_1 \cup \mathbf{P}_2 \cup \dots \cup \mathbf{P}_n \subset \mathbb{E}^d$*), then*

$$\sum_{i=1}^{n} w(\mathbf{P}_i) \geq w(\mathbf{C}).$$

In [18], Bang raised the following stronger version of Tarski's plank problem called the affine plank problem. We phrase it via the following definition. Let $\mathbf{C}$ be a convex body and let $\mathbf{P}$ be a plank with boundary hyperplanes parallel to the hyperplane H in $\mathbb{E}^d$. We define the $\mathbf{C}$-*width* of the plank $\mathbf{P}$ as $\frac{w(\mathbf{P})}{w(\mathbf{C},H)}$ and label it $w_{\mathbf{C}}(\mathbf{P})$. (This notion was introduced by Bang [18] under the name "relative width".)

K. Bezdek, *Classical Topics in Discrete Geometry*, CMS Books in Mathematics, DOI 10.1007/978-1-4419-0600-7_4, © Springer Science+Business Media, LLC 2010

Conjecture 4.1.2 *If the convex body* **C** *is covered by the planks* $\mathbf{P}_1, \mathbf{P}_2, \ldots,$ $\mathbf{P}_n$ *in* $\mathbb{E}^d, d \geq 2$, *then*

$$\sum_{i=1}^{n} w_{\mathbf{C}}(\mathbf{P}_i) \geq 1.$$

The special case of Conjecture 4.1.2, when the convex body to be covered is centrally symmetric, has been proved by Ball in [12]. Thus, the following is Ball's plank theorem.

Theorem 4.1.3 *If the centrally symmetric convex body* **C** *is covered by the planks* $\mathbf{P}_1, \mathbf{P}_2, \ldots, \mathbf{P}_n$ *in* $\mathbb{E}^d, d \geq 2$, *then*

$$\sum_{i=1}^{n} w_{\mathbf{C}}(\mathbf{P}_i) \geq 1.$$

From the point of view of discrete geometry it seems natural to mention that after proving Theorem 4.1.3 Ball [13] used Bang's proof of Theorem 4.1.1 to derive a new argument for an improvement of the Davenport–Rogers lower bound on the density of economical sphere lattice packings.

It was Alexander [3] who noticed that Conjecture 4.1.2 is equivalent to the following generalization of a problem of Davenport.

Conjecture 4.1.4 *If a convex body* **C** *in* $\mathbb{E}^d, d \geq 2$ *is sliced by* $n - 1$ *hyperplane cuts, then there exists a piece that covers a translate of* $\frac{1}{n}\mathbf{C}$.

We note that the paper [33] of A. Bezdek and the author proves Conjecture 4.1.4 for successive hyperplane cuts (i.e., for hyperplane cuts when each cut divides one piece). Also, the same paper ([33]) introduced two additional equivalent versions of Conjecture 4.1.2. As they seem to be of independent interest we recall them following the terminology used in [33].

Let **C** and **K** be convex bodies in $\mathbb{E}^d$ and let H be a hyperplane of $\mathbb{E}^d$. The **C**-*width of* **K** *parallel to* H is denoted by $w_{\mathbf{C}}(\mathbf{K}, H)$ and is defined as $\frac{w(\mathbf{K}, H)}{w(\mathbf{C}, H)}$. The *minimal* **C**-*width of* **K** is denoted by $w_{\mathbf{C}}(\mathbf{K})$ and is defined as the minimum of $w_{\mathbf{C}}(\mathbf{K}, H)$, where the minimum is taken over all possible hyperplanes H of $\mathbb{E}^d$. Recall that the inradius of **K** is the radius of the largest ball contained in **K**. It is quite natural then to introduce the **C**-*inradius of* **K** as the factor of the largest (positively) homothetic copy of **C**, a translate of which is contained in **K**. We need to do one more step to introduce the so-called successive **C**-inradii of **K** as follows. Let r be the **C**-inradius of **K**. For any $0 < \rho \leq r$ let the ρ**C**-*rounded body of* **K** be denoted by $\mathbf{K}^{\rho\mathbf{C}}$ and be defined as the union of all translates of $\rho\mathbf{C}$ that are covered by **K**. Now, take a fixed integer $n \geq 1$. On the one hand, if $\rho > 0$ is sufficiently small, then $w_{\mathbf{C}}(\mathbf{K}^{\rho\mathbf{C}}) > n\rho$. On the other hand, $w_{\mathbf{C}}(\mathbf{K}^{r\mathbf{C}}) = r \leq nr$. As $w_{\mathbf{C}}(\mathbf{K}^{\rho\mathbf{C}})$ is a decreasing continuous function of $\rho > 0$ and $n\rho$ is a strictly increasing continuous function of ρ, there exists a uniquely determined $\rho > 0$ such that

$$w_{\mathbf{C}}(\mathbf{K}^{\rho\mathbf{C}}) = n\rho.$$

This uniquely determined ρ is called the *nth successive* $\mathbf{C}$-*inradius of* $\mathbf{K}$ and is denoted by $r_{\mathbf{C}}(\mathbf{K}, n)$. Notice that $r_{\mathbf{C}}(\mathbf{K}, 1) = r$. Now, the two equivalent versions of Conjecture 4.1.2 and Conjecture 4.1.4 introduced in [33] can be phrased as follows.

Conjecture 4.1.5 *If a convex body* $\mathbf{K}$ *in* $\mathbb{E}^d, d \geq 2$ *is covered by the planks* $\mathbf{P}_1, \mathbf{P}_2, \ldots, \mathbf{P}_n$, *then* $\sum_{i=1}^{n} w_{\mathbf{C}}(\mathbf{P}_i) \geq w_{\mathbf{C}}(\mathbf{K})$ *for any convex body* $\mathbf{C}$ *in* $\mathbb{E}^d$.

Conjecture 4.1.6 *Let* $\mathbf{K}$ *and* $\mathbf{C}$ *be convex bodies in* $\mathbb{E}^d, d \geq 2$. *If* $\mathbf{K}$ *is sliced by* $n - 1$ *hyperplanes, then the minimum of the greatest* $\mathbf{C}$-*inradius of the pieces is equal to the nth successive* $\mathbf{C}$-*inradius of* $\mathbf{K}$; *that is, it is* $r_{\mathbf{C}}(\mathbf{K}, n)$.

A. Bezdek and the author [33] proved the following theorem that (under the condition that $\mathbf{C}$ is a ball) answers a question raised by Conway ([32]) as well as proves Conjecture 4.1.6 for successive hyperplane cuts.

Theorem 4.1.7 *Let* $\mathbf{K}$ *and* $\mathbf{C}$ *be convex bodies in* $\mathbb{E}^d$, $d \geq 2$. *If* $\mathbf{K}$ *is sliced into* n *pieces by* $n - 1$ *successive hyperplane cuts (i.e., when each cut divides one piece), then the minimum of the greatest* $\mathbf{C}$-*inradius of the pieces is the nth successive* $\mathbf{C}$-*inradius of* $\mathbf{K}$ *(i.e.,* $r_{\mathbf{C}}(\mathbf{K}, n)$). *An optimal partition is achieved by* $n - 1$ *parallel hyperplane cuts equally spaced along the minimal* $\mathbf{C}$-*width of the* $r_{\mathbf{C}}(\mathbf{K}, n)\mathbf{C}$-*rounded body of* $\mathbf{K}$.

4.2 Covering Convex Bodies by Cylinders

In his paper [18], Bang, by describing a concrete example and writing that it may be extremal, proposes investigating a quite challenging question that can be phrased as follows.

Problem 4.2.1 *Prove or disprove that the sum of the base areas of finitely many cylinders covering a 3-dimensional convex body is at least half of the minimum area 2-dimensional projection of the body.*

If true, then the estimate of Problem 4.2.1 is a sharp one due to a covering of a regular tetrahedron by two cylinders described in [18]. A very recent paper of the author and Litvak ([71]) investigates Problem 4.2.1 as well as its higher-dimensional analogue. Their main result can be summarized as follows.

Given $0 < k < d$ define a k-codimensional cylinder $\mathbf{C}$ in $\mathbb{E}^d$ as a set which can be presented in the form $\mathbf{C} = H + B$, where H is a k-dimensional linear subspace of $\mathbb{E}^d$ and B is a measurable set (called the base) in the orthogonal complement $H^{\perp}$ of H. For a given convex body $\mathbf{K}$ and a k-codimensional cylinder $\mathbf{C} = H + B$ we define the cross-sectional volume $\text{crv}_{\mathbf{K}}(\mathbf{C})$ of $\mathbf{C}$ with respect to $\mathbf{K}$ as follows,

$$\mathrm{crv}_{\mathbf{K}}(\mathbf{C}) := \frac{\mathrm{vol}_{d-k}(\mathbf{C} \cap H^{\perp})}{\mathrm{vol}_{d-k}(P_{H^{\perp}}\mathbf{K})} = \frac{\mathrm{vol}_{d-k}(P_{H^{\perp}}\mathbf{C})}{\mathrm{vol}_{d-k}(P_{H^{\perp}}\mathbf{K})} = \frac{\mathrm{vol}_{d-k}(B)}{\mathrm{vol}_{d-k}(P_{H^{\perp}}\mathbf{K})},$$

where $P_{H^{\perp}} : \mathbb{E}^d \to H^{\perp}$ denotes the orthogonal projection of $\mathbb{E}^d$ onto $H^{\perp}$. Notice that for every invertible affine map $T : \mathbb{E}^d \to \mathbb{E}^d$ one has $\mathrm{crv}_{\mathbf{K}}(\mathbf{C}) = \mathrm{crv}_{T\mathbf{K}}(T\mathbf{C})$. The following theorem is proved in [71].

Theorem 4.2.2 *Let $\mathbf{K}$ be a convex body in $\mathbb{E}^d$. Let $\mathbf{C}_1, \dots, \mathbf{C}_N$ be k-codimensional cylinders in $\mathbb{E}^d$, $0 < k < d$ such that $\mathbf{K} \subset \bigcup_{i=1}^{N} \mathbf{C}_i$. Then*

$$\sum_{i=1}^{N} \mathrm{crv}_{\mathbf{K}}(\mathbf{C}_i) \geq \frac{1}{\binom{d}{k}}.$$

Moreover, if $\mathbf{K}$ is an ellipsoid and $\mathbf{C}_1, \dots, \mathbf{C}_N$ are 1-codimensional cylinders in $\mathbb{E}^d$ such that $\mathbf{K} \subset \bigcup_{i=1}^{N} \mathbf{C}_i$, then

$$\sum_{i=1}^{N} \mathrm{crv}_{\mathbf{K}}(\mathbf{C}_i) \geq 1.$$

The case $k = d - 1$ of Theorem 4.2.2 corresponds to Conjecture 4.1.2, that is, to the affine plank problem. Theorem 4.2.2 for $k = d - 1$ implies the lower bound $1/d$ that can be somewhat further improved (for more details see [71]).

As an immediate corollary of Theorem 4.2.2 we get the following estimate for Problem 4.2.1.

Corollary 4.2.3 *The sum of the base areas of finitely many (1-codimensional) cylinders covering a 3-dimensional convex body is always at least one third of the minimum area 2-dimensional projection of the body.*

Also, note that the inequality of Theorem 4.2.2 on covering ellipsoids by 1-codimensional cylinders is best possible. By looking at this result from the point of view of k-codimensional cylinders we are led to ask the following quite natural question. Unfortunately, despite its elementary character it is still open.

Problem 4.2.4 *Let $0 < c(d, k) \leq 1$ denote the largest real number with the property that if $\mathbf{K}$ is an ellipsoid and $\mathbf{C}_1, \dots, \mathbf{C}_N$ are k-codimensional cylinders in $\mathbb{E}^d$, $1 \leq k \leq d - 1$ such that $\mathbf{K} \subset \bigcup_{i=1}^{N} \mathbf{C}_i$, then $\sum_{i=1}^{N} \mathrm{crv}_{\mathbf{K}}(\mathbf{C}_i) \geq c(d, k)$. Determine $c(d, k)$ for given d and k.*

On the one hand, Theorems 4.1.1 and 4.2.2 imply that $c(d, d - 1) = 1$ and $c(d, 1) = 1$; moreover, $c(d, k) \geq \frac{1}{\binom{d}{k}}$. On the other hand, a clever construction due to Kadets [174] shows that if $d - k \geq 3$ is a fixed integer, then $\lim_{d \to \infty} c(d, k) = 0$. Thus the following as a subquestion of Problem 4.2.4 seems to be open as well.

Problem 4.2.5 *Prove or disprove the existence of a universal constant $c > 0$ (independent of d) with the property that if $\mathbf{B}^d$ denotes the unit ball centered at the origin $\mathbf{o}$ in $\mathbb{E}^d$ and $\mathbf{C}_1, \ldots, \mathbf{C}_N$ are $(d-2)$-codimensional cylinders in $\mathbb{E}^d$ such that $\mathbf{B}^d \subset \bigcup_{i=1}^{N} \mathbf{C}_i$, then the sum of the 2-dimensional base areas of $\mathbf{C}_1, \ldots, \mathbf{C}_N$ is at least c.*

4.3 Covering Lattice Points by Hyperplanes

In their paper [51], the author and Hausel established the following discrete version of Tarski's plank problem.

Recall that the lattice width of a convex body $\mathbf{K}$ in $\mathbb{E}^d$ is defined as

$$w(\mathbf{K}, \mathbb{Z}^d) = \min\Big\{ \max_{\mathbf{x} \in \mathbf{K}} \langle \mathbf{x}, \mathbf{y} \rangle - \min_{\mathbf{x} \in \mathbf{K}} \langle \mathbf{x}, \mathbf{y} \rangle \mid \mathbf{y} \in \mathbb{Z}^d, \mathbf{y} \neq \mathbf{o} \Big\},$$

where $\mathbb{Z}^d$ denotes the integer lattice of $\mathbb{E}^d$. It is well known that if $\mathbf{y} \in \mathbb{Z}^d$, $\mathbf{y} \neq \mathbf{o}$ is chosen such that $\lambda \mathbf{y} \notin \mathbb{Z}^d$ for any $0 < \lambda < 1$ (i.e., $\mathbf{y}$ is a primitive integer point), then

$$\max_{\mathbf{x} \in \mathbf{K}} \langle \mathbf{x}, \mathbf{y} \rangle - \min_{\mathbf{x} \in \mathbf{K}} \langle \mathbf{x}, \mathbf{y} \rangle$$

is equal to the Euclidean width of $\mathbf{K}$ in the direction $\mathbf{y}$ divided by the Euclidean distance between two consecutive lattice hyperplanes of $\mathbb{Z}^d$ that are orthogonal to $\mathbf{y}$. Thus if $\mathbf{K}$ is the convex hull of finitely many points of $\mathbb{Z}^d$, then

$$\max_{\mathbf{x} \in \mathbf{K}} \langle \mathbf{x}, \mathbf{y} \rangle - \min_{\mathbf{x} \in \mathbf{K}} \langle \mathbf{x}, \mathbf{y} \rangle$$

is an integer namely, it is less by one than the number of lattice hyperplanes of $\mathbb{Z}^d$ that intersect $\mathbf{K}$ and are orthogonal to $\mathbf{y}$. Now, we are ready to state the following conjecture of the author and Hausel ([51]).

Conjecture 4.3.1 *Let $\mathbf{K}$ be a convex body in $\mathbb{E}^d$. Let $H_1, \ldots, H_N$ be hyperplanes in $\mathbb{E}^d$ such that*

$$\mathbf{K} \cap \mathbb{Z}^d \subset \bigcup_{i=1}^{N} H_i.$$

Then

$$N \geq w(\mathbf{K}, \mathbb{Z}^d) - d.$$

Properly translated copies of cross-polytopes, described in [51], show that if true, then the above inequality is best possible.

The special case, when $N = 0$, is of independent interest. (In particular, this case seems to be "responsible" for the term d in the inequality of Conjecture 4.3.1.) Namely, it seems reasonable to conjecture (see also [16]) that if $\mathbf{K}$ is an integer point free convex body in $\mathbb{E}^d$, then $w(\mathbf{K}, \mathbb{Z}^d) \leq d$. On the one hand, this has been proved by Banaszczyk [15] for ellipsoids. On the other hand, for general convex bodies containing no integer points, Banaszczyk, Litvak, Pajor, and Szarek [16] have proved the inequality $w(\mathbf{K}, \mathbb{Z}^d) \le C d^{\frac{3}{2}}$, where C is an absolute positive constant. This improves an earlier result of Kannan and Lovász [177].

Although Conjecture 4.3.1 is still open we have the following partial results which were recently published. Improving the estimates of [51], Talata [238] has succeeded in deriving a proof of the following inequality.

Theorem 4.3.2 *Let* $\mathbf{K}$ *be a convex body in* $\mathbb{E}^d$. *Let* $H_1, \ldots, H_N$ *be hyperplanes in* $\mathbb{E}^d$ *such that*

$$\mathbf{K} \cap \mathbb{Z}^d \subset \bigcup_{i=1}^N H_i.$$

Then

$$N \ge c \frac{w(\mathbf{K}, \mathbb{Z}^d)}{d} - d,$$

where c *is an absolute positive constant.*

In the paper [71], the author and Litvak have shown that the plank theorem of Ball [12] implies a slight improvement on the above inequality for centrally symmetric convex bodies whose lattice width is at most quadratic in dimension. (Actually, this approach is different from Talata's technique and can lead to a somewhat even stronger inequality in terms of the relevant basic measure of the given convex body. For more details on this we refer the interested reader to [71].)

Theorem 4.3.3 *Let* $\mathbf{K}$ *be a centrally symmetric convex body in* $\mathbb{E}^d$. *Let* $H_1, \ldots, H_N$ *be hyperplanes in* $\mathbb{E}^d$ *such that*

$$\mathbf{K} \cap \mathbb{Z}^d \subset \bigcup_{i=1}^N H_i.$$

Then

$$N \ge c \frac{w(\mathbf{K}, \mathbb{Z}^d)}{d \ln(d+1)},$$

where c *is an absolute positive constant.*

Motivated by Conjecture 4.3.1 and by a conjecture of Corzatt [109] (according to which if in the plane the integer points of a convex domain can be covered by N lines, then those integer points can also be covered by N lines having at most four different slopes), Brass, Moser, and Pach [96] have raised the following related question.

Problem 4.3.4 *For every positive integer* d *find the smallest constant* $c(d)$ *such that if the integer points of a convex body in* $\mathbb{E}^d$ *can be covered by* N *hyperplanes, then those integer points can also be covered by* $c(d)N$ *parallel hyperplanes.*

Theorem 4.3.2 implies that $c(d) \leq c\,d^2$ for convex bodies in general and for centrally symmetric convex bodies Theorem 4.3.3 yields the somewhat better upper bound $c\,d\ln(d+1)$. As a last note we mention that the problem of finding good estimates for the constants of Theorems 4.3.2 and 4.3.3 is an interesting open question as well.

4.4 On Some Strengthenings of the Plank Theorems of Ball and Bang

Recall that Ball ([12]) generalized the plank theorem of Bang ([17], [18]) for coverings of balls by planks in Banach spaces (where planks are defined with the help of linear functionals instead of inner product). This theorem was further strengthened by Kadets [175] for real Hilbert spaces as follows. Let $\mathbf{C}$ be a closed convex subset with non-empty interior in the real Hilbert space $\mathbb{H}$ (finite or infinite dimensional). We call $\mathbf{C}$ a *convex body* of $\mathbb{H}$. Then let $\mathrm{r}(\mathbf{C})$ denote the supremum of the radii of the balls contained in $\mathbf{C}$. (One may call $\mathrm{r}(\mathbf{C})$ the *inradius* of $\mathbf{C}$.) Planks and their widths in $\mathbb{H}$ are defined with the help of the inner product of $\mathbb{H}$ in the usual way. Thus, if $\mathbf{C}$ is a convex body in $\mathbb{H}$ and $\mathbf{P}$ is a plank of $\mathbb{H}$, then the width $\mathrm{w}(\mathbf{P})$ of $\mathbf{P}$ is always at least as large as $2\mathrm{r}(\mathbf{C} \cap \mathbf{P})$. Now, the main result of [175] is the following.

Theorem 4.4.1 *Let the ball $\mathbf{B}$ of the real Hilbert space $\mathbb{H}$ be covered by the convex bodies $\mathbf{C}_1, \mathbf{C}_2, \ldots, \mathbf{C}_n$ in $\mathbb{H}$. Then*

$$\sum_{i=1}^{n} \mathrm{r}(\mathbf{C}_i \cap \mathbf{B}) \geq \mathrm{r}(\mathbf{B}).$$

We note that an independent proof of the 2-dimensional Euclidean case of Theorem 4.4.1 can be found in [35]. Kadets ([175]) proposes to investigate the analogue of Theorem 4.4.1 in Banach spaces. Thus, an affirmative answer to the following problem would improve the plank theorem of Ball.

Problem 4.4.2 *Let the ball $\mathbf{B}$ be covered by the convex bodies $\mathbf{C}_1, \mathbf{C}_2, \ldots, \mathbf{C}_n$ in an arbitrary Banach space. Prove or disprove that*

$$\sum_{i=1}^{n} \mathrm{r}(\mathbf{C}_i \cap \mathbf{B}) \geq \mathrm{r}(\mathbf{B}).$$

In order to complete the picture on plank-type results in spaces other than Euclidean we mention the statement below, proved by Schneider and the author [74]. It is an extension of Theorem 4.4.1 for coverings of large balls in spherical spaces. Recall that $\mathbb{S}^d$ stands for the d-dimensional unit sphere in $(d+1)$-dimensional Euclidean space $\mathbb{E}^{d+1}, d \geq 2$. A *spherically convex body* is a closed, spherically convex subset $\mathbf{K}$ of $\mathbb{S}^d$ with interior points and lying in some closed hemisphere, thus, the intersection of $\mathbb{S}^d$ with a $(d+1)$-dimensional closed convex cone of $\mathbb{E}^{d+1}$ different from $\mathbb{E}^{d+1}$. The *inradius* $r(\mathbf{K})$ of $\mathbf{K}$ is the spherical radius of the largest spherical ball contained in $\mathbf{K}$. Also, recall that a *lune* in $\mathbb{S}^d$ is the d-dimensional intersection of $\mathbb{S}^d$ with two closed halfspaces of $\mathbb{E}^{d+1}$ with the origin $\mathbf{o}$ in their boundaries. The intersection of the boundaries (or any $(d-1)$-dimensional subspace in that intersection, if the two subspaces are identical) is called the *ridge* of the lune. Evidently, the inradius of a lune is half the interior angle between the two defining hyperplanes.

Theorem 4.4.3 *If the spherically convex bodies* $\mathbf{K}_1, \ldots, \mathbf{K}_n$ *cover the spherical ball* $\mathbf{B}$ *of radius* $r(\mathbf{B}) \geq \frac{\pi}{2}$ *in* $\mathbb{S}^d, d \geq 2$, *then*

$$\sum_{i=1}^{n} r(\mathbf{K}_i) \geq r(\mathbf{B}).$$

For $r(\mathbf{B}) = \frac{\pi}{2}$ *the stronger inequality* $\sum_{i=1}^{n} r(\mathbf{K}_i \cap \mathbf{B}) \geq r(\mathbf{B})$ *holds. Moreover, equality for* $r(\mathbf{B}) = \pi$ *or* $r(\mathbf{B}) = \frac{\pi}{2}$ *holds if and only if* $\mathbf{K}_1, \ldots, \mathbf{K}_n$ *are lunes with common ridge which have pairwise no common interior points.*

Theorem 4.4.3 is a consequence of the following result proved by Schneider and the author in [74]. Recall that $\mathrm{Svol}_d(\ldots)$ denotes the spherical Lebesgue measure on $\mathbb{S}^d$, and recall that $(d+1)\omega_{d+1} = \mathrm{Svol}_d(\mathbb{S}^d)$.

Theorem 4.4.4 *If* $\mathbf{K}$ *is a spherically convex body in* $\mathbb{S}^d, d \geq 2$, *then*

$$\mathrm{Svol}_d(\mathbf{K}) \leq \frac{(d+1)\omega_{d+1}}{\pi} r(\mathbf{K}).$$

Equality holds if and only if $\mathbf{K}$ *is a lune.*

Indeed, Theorem 4.4.4 implies Theorem 4.4.3 as follows. If $\mathbf{B} = \mathbb{S}^d$; that is, the spherically convex bodies $\mathbf{K}_1, \ldots, \mathbf{K}_n$ cover $\mathbb{S}^d$, then

$$(d+1)\omega_{d+1} \leq \sum_{i=1}^{n} \mathrm{Svol}_d(\mathbf{K}_i) \leq \frac{(d+1)\omega_{d+1}}{\pi} \sum_{i=1}^{n} r(\mathbf{K}_i),$$

and the stated inequality follows. In general, when $\mathbf{B}$ is different from $\mathbb{S}^d$, let $\mathbf{B}' \subset \mathbb{S}^d$ be the spherical ball of radius $\pi - r(\mathbf{B})$ centered at the point antipodal to the center of $\mathbf{B}$. As the spherically convex bodies $\mathbf{B}', \mathbf{K}_1, \ldots, \mathbf{K}_n$ cover $\mathbb{S}^d$, the inequality just proved shows that

$$\pi - r(\mathbf{B}) + \sum_{i=1}^{n} r(\mathbf{K}_i) \geq \pi,$$

and the stated inequality follows. If $r(\mathbf{B}) = \frac{\pi}{2}$, then $\mathbf{K}_1 \cap \mathbf{B}, \ldots, \mathbf{K}_n \cap \mathbf{B}$ are spherically convex bodies and as $\mathbf{B}', \mathbf{K}_1 \cap \mathbf{B}, \ldots, \mathbf{K}_n \cap \mathbf{B}$ cover $\mathbb{S}^d$, the stronger inequality follows. The assertion about the equality sign for the case when $r(\mathbf{B}) = \pi$ or $r(\mathbf{B}) = \frac{\pi}{2}$ follows easily.

4.5 On Partial Coverings by Planks: Bang's Theorem Revisited

The following variant of Tarski's plank problem was introduced very recently by the author in [73]: let $\mathbf{C}$ be a convex body of minimal width $w > 0$ in $\mathbb{E}^d$. Moreover, let $w_1 > 0, w_2 > 0, \ldots, w_n > 0$ be given with $w_1 + w_2 + \cdots + w_n < w$. Then find the arrangement of n planks say, of $\mathbf{P}_1, \mathbf{P}_2, \ldots, \mathbf{P}_n$, of width $w_1, w_2, \ldots, w_n$ in $\mathbb{E}^d$ such that their union covers the largest volume subset of $\mathbf{C}$, that is, for which $\mathrm{vol}_d((\mathbf{P}_1 \cup \mathbf{P}_2 \cup \cdots \cup \mathbf{P}_n) \cap \mathbf{C})$ is as large as possible. As the following special case is the most striking form of the above problem, we are putting it forward as the main question of this section.

Problem 4.5.1 *Let $\mathbf{B}^d$ denote the unit ball centered at the origin $\mathbf{o}$ in $\mathbb{E}^d$. Moreover, let $w_1, w_2, \ldots, w_n$ be positive real numbers satisfying the inequality $w_1 + w_2 + \cdots + w_n < 2$. Then prove or disprove that the union of the planks $\mathbf{P}_1, \mathbf{P}_2, \ldots, \mathbf{P}_n$ of width $w_1, w_2, \ldots, w_n$ in $\mathbb{E}^d$ covers the largest volume subset of $\mathbf{B}^d$ if and only if $\mathbf{P}_1 \cup \mathbf{P}_2 \cup \cdots \cup \mathbf{P}_n$ is a plank of width $w_1 + w_2 + \cdots + w_n$ with $\mathbf{o}$ as a center of symmetry.*

Clearly, there is an affirmative answer to Problem 4.5.1 for $n = 1$. Also, we note that it would not come as a surprise to us if it turned out that the answer to Problem 4.5.1 is positive in proper low dimensions and negative in (sufficiently) high dimensions. The following partial results have been obtained in [73].

Theorem 4.5.2 *Let $w_1, w_2, \ldots, w_n$ be positive real numbers satisfying the inequality $w_1 + w_2 + \cdots + w_n < 2$. Then the union of the planks $\mathbf{P}_1, \mathbf{P}_2, \ldots, \mathbf{P}_n$ of width $w_1, w_2, \ldots, w_n$ in $\mathbb{E}^3$ covers the largest volume subset of $\mathbf{B}^3$ if and only if $\mathbf{P}_1 \cup \mathbf{P}_2 \cup \cdots \cup \mathbf{P}_n$ is a plank of width $w_1 + w_2 + \cdots + w_n$ with $\mathbf{o}$ as a center of symmetry.*

Corollary 4.5.3 *If $\mathbf{P}_1, \mathbf{P}_2$, and $\mathbf{P}_3$ are planks in $\mathbb{E}^d$, $d \geq 3$ of widths w_1, w_2, and w_3 satisfying $0 < w_1 + w_2 + w_3 < 2$, then $\mathbf{P}_1 \cup \mathbf{P}_2 \cup \mathbf{P}_3$ covers the largest volume subset of $\mathbf{B}^d$ if and only if $\mathbf{P}_1 \cup \mathbf{P}_2 \cup \mathbf{P}_3$ is a plank of width $w_1 + w_2 + w_3$ having $\mathbf{o}$ as a center of symmetry.*

The following estimate of [73] can be derived from Bang's paper [18]. In order to state it properly we introduce two definitions.

Definition 4.5.4 *Let $\mathbf{C}$ be a convex body in $\mathbb{E}^d$ and let m be a positive integer. Then let $\mathcal{T}_{\mathbf{C},d}^m$ denote the family of all sets in $\mathbb{E}^d$ that can be obtained as the intersection of at most m translates of $\mathbf{C}$ in $\mathbb{E}^d$.*

Definition 4.5.5 *Let $\mathbf{C}$ be a convex body of minimal width $w > 0$ in $\mathbb{E}^d$ and let $0 < x \leq w$ be given. Then for any non-negative integer n let*

$$v_d(\mathbf{C}, x, n) := \min\{\mathrm{vol}_d(\mathbf{Q}) \mid \mathbf{Q} \in \mathcal{T}_{\mathbf{C},d}^{2^n} \text{ and } w(\mathbf{Q}) \geq x \}.$$

Now, we are ready to state the theorem which although it was not published by Bang in [18], follows from his proof of Tarski's plank conjecture.

Theorem 4.5.6 *Let* $\mathbf{C}$ *be a convex body of minimal width* $w > 0$ *in* $\mathbb{E}^d$. *Moreover, let* $\mathbf{P}_1, \mathbf{P}_2, \ldots, \mathbf{P}_n$ *be planks of width* $w_1, w_2, \ldots, w_n$ *in* $\mathbb{E}^d$ *with* $w_0 = w_1 + w_2 + \cdots + w_n < w$. *Then*

$$\mathrm{vol}_d(\mathbf{C} \setminus (\mathbf{P}_1 \cup \mathbf{P}_2 \cup \cdots \cup \mathbf{P}_n)) \geq \mathrm{v}_d(\mathbf{C}, w - w_0, n);$$

that is,

$$\mathrm{vol}_d((\mathbf{P}_1 \cup \mathbf{P}_2 \cup \cdots \cup \mathbf{P}_n) \cap \mathbf{C}) \leq \mathrm{vol}_d(\mathbf{C}) - \mathrm{v}_d(\mathbf{C}, w - w_0, n).$$

Clearly, the first inequality above implies (via an indirect argument) that if the planks $\mathbf{P}_1, \mathbf{P}_2, \ldots, \mathbf{P}_n$ of width $w_1, w_2, \ldots, w_n$ cover the convex body $\mathbf{C}$ in $\mathbb{E}^d$, then $w_1 + w_2 + \cdots + w_n \geq w$. Also, as an additional observation from [73] we mention the following statement, that can be derived from Theorem 4.5.6 in a straightforward way and, on the other hand, represents the only case when the estimate in Theorem 4.5.6 is sharp.

Corollary 4.5.7 *Let* $\mathbf{T}$ *be an arbitrary triangle of minimal width (i.e., of minimal height)* $w > 0$ *in* $\mathbb{E}^2$. *Moreover, let* $w_1, w_2, \ldots, w_n$ *be positive real numbers satisfying the inequality* $w_1 + w_2 + \cdots + w_n < w$. *Then the union of the planks* $\mathbf{P}_1, \mathbf{P}_2, \ldots, \mathbf{P}_n$ *of width* $w_1, w_2, \ldots, w_n$ *in* $\mathbb{E}^2$ *covers the largest area subset of* $\mathbf{T}$ *if* $\mathbf{P}_1 \cup \mathbf{P}_2 \cup \cdots \cup \mathbf{P}_n$ *is a plank of width* $w_1 + w_2 + \cdots + w_n$ *sitting on the side of* $\mathbf{T}$ *with height* w.

It was observed by the author in [73] that there is an implicit connection between problem 4.5.1 and the well-known Blaschke–Lebesgue problem, which is generated by Theorem 4.5.6. The details are as follows.

First, recall that the *Blaschke–Lebesgue problem* is about finding the minimum volume convex body of constant width $w > 0$ in $\mathbb{E}^d$. In particular, the Blaschke–Lebesgue theorem states that among all convex domains of constant width w, the Reuleaux triangle of width w has the smallest area, namely $\frac{1}{2}(\pi - \sqrt{3})w^2$. Blaschke [76] and Lebesgue [188] were the first to show this and the succeeding decades have seen other works published on different proofs of that theorem. For a most recent new proof, and for a survey on the state of the art of different proofs of the Blaschke–Lebesgue theorem, see the elegant paper of Harrell [167]. Here we note that the Blaschke–Lebesgue problem is unsolved in three and more dimensions. Even finding the 3-dimensional set of least volume presents formidable difficulties. On the one hand, Chakerian [101] proved that any convex body of constant width 1 in $\mathbb{E}^3$ has volume at least $\frac{\pi(3\sqrt{6}-7)}{3} = 0.365\ldots$. On the other hand, it has been conjectured by Bonnesen and Fenchel [85] that Meissner's 3-dimensional generalizations of the Reuleaux triangle of volume $\pi(\frac{2}{3} - \frac{1}{4}\sqrt{3}\arccos(\frac{1}{3})) = 0.420\ldots$ are the only extramal sets in $\mathbb{E}^3$.

For our purposes it is useful to introduce the notation $\mathbf{K}_{BL}^{w,d}$ (resp., $\overline{\mathbf{K}}_{BL}^{w,d}$) for a convex body of constant width w in $\mathbb{E}^d$ having minimum volume (resp., surface volume). One may call $\mathbf{K}_{BL}^{w,d}$ (resp., $\overline{\mathbf{K}}_{BL}^{w,d}$) a Blaschke–Lebesgue-type convex body with respect to volume (resp., surface volume). Note that for $d = 2, 3$ one may choose $\mathbf{K}_{BL}^{w,d} = \overline{\mathbf{K}}_{BL}^{w,d}$, however, this is likely not to happen for $d \geq 4$. (For more details on this see [101].) As an important note we mention that Schramm [227] has proved the inequality

$$\mathrm{vol}_d(\mathbf{K}_{BL}^{w,d}) \geq \left(\sqrt{3 + \frac{2}{d+1}} - 1\right)^d \left(\frac{w}{2}\right)^d \mathrm{vol}_d(\mathbf{B}^d),$$

which gives the best lower bound for all $d > 4$. By observing that the orthogonal projection of a convex body of constant width w in $\mathbb{E}^d$ onto any hyperplane of $\mathbb{E}^d$ is a $(d-1)$-dimensional convex body of constant width w one obtains from the previous inequality of Schramm the following one,

$$\mathrm{svol}_{d-1}(\mathrm{bd}(\overline{\mathbf{K}}_{BL}^{w,d})) \geq d\left(\sqrt{3 + \frac{2}{d}} - 1\right)^{d-1} \left(\frac{w}{2}\right)^{d-1} \mathrm{vol}_d(\mathbf{B}^d).$$

Second, let us recall that if X is a finite (point) set lying in the interior of a unit ball in $\mathbb{E}^d$, then the intersection of the (closed) unit balls of $\mathbb{E}^d$ centered at the points of X is called a ball-polyhedron and it is denoted by $\mathbf{B}[X]$. (For an extensive list of properties of ball-polyhedra see the recent paper [69].) Of course, it also makes sense to introduce $\mathbf{B}[X]$ for sets X that are not finite but in those cases we get sets that are typically not ball-polyhedra.

Now, we are ready to state our theorem.

Theorem 4.5.8 *Let $\mathbf{B}[X] \subset \mathbb{E}^d$ be a ball-polyhedron of minimal width x with $1 \leq x < 2$. Then*

$$\mathrm{vol}_d(\mathbf{B}[X]) \geq \mathrm{vol}_d(\mathbf{K}_{BL}^{2-x,d}) + \mathrm{svol}_{d-1}(\mathrm{bd}(\overline{\mathbf{K}}_{BL}^{2-x,d}))(x-1) + \mathrm{vol}_d(\mathbf{B}^d)(x-1)^d.$$

Thus, Theorem 4.5.6 and Theorem 4.5.8 imply the following immediate estimate.

Corollary 4.5.9 *Let $\mathbf{B}^d$ denote the unit ball centered at the origin $\mathbf{o}$ in $\mathbb{E}^d$, $d \geq 2$. Moreover, let $\mathbf{P}_1, \mathbf{P}_2, \ldots, \mathbf{P}_n$ be planks of width $w_1, w_2, \ldots, w_n$ in $\mathbb{E}^d$ with $w_0 = w_1 + w_2 + \cdots + w_n \leq 1$. Then*

$$\mathrm{vol}_d((\mathbf{P}_1 \cup \mathbf{P}_2 \cup \cdots \cup \mathbf{P}_n) \cap \mathbf{B}^d) \leq \mathrm{vol}_d(\mathbf{B}^d) - v_d(\mathbf{B}^d, 2 - w_0, n)$$

$$\leq (1 - (1-w_0)^d)\mathrm{vol}_d(\mathbf{B}^d) - \mathrm{vol}_d(\mathbf{K}_{BL}^{w_0,d}) - \mathrm{svol}_{d-1}(\mathrm{bd}(\overline{\mathbf{K}}_{BL}^{w_0,d}))(1 - w_0).$$

5

On the Volume of Finite Arrangements of Spheres

5.1 The Conjecture of Kneser and Poulsen

Recall that $\| \ldots \|$ denotes the standard Euclidean norm of the d-dimensional Euclidean space $\mathbb{E}^d$. So, if $\mathbf{p}_i, \mathbf{p}_j$ are two points in $\mathbb{E}^d$, then $\|\mathbf{p}_i - \mathbf{p}_j\|$ denotes the Euclidean distance between them. It is convenient to denote the (finite) point configuration consisting of the points $\mathbf{p}_1, \mathbf{p}_2, \ldots, \mathbf{p}_N$ in $\mathbb{E}^d$ by $\mathbf{p} = (\mathbf{p}_1, \mathbf{p}_2, \ldots, \mathbf{p}_N)$. Now, if $\mathbf{p} = (\mathbf{p}_1, \mathbf{p}_2, \ldots, \mathbf{p}_N)$ and $\mathbf{q} = (\mathbf{q}_1, \mathbf{q}_2, \ldots, \mathbf{q}_N)$ are two configurations of N points in $\mathbb{E}^d$ such that for all $1 \leq i < j \leq N$ the inequality $\|\mathbf{q}_i - \mathbf{q}_j\| \leq \|\mathbf{p}_i - \mathbf{p}_j\|$ holds, then we say that $\mathbf{q}$ is a *contraction* of $\mathbf{p}$. If $\mathbf{q}$ is a contraction of $\mathbf{p}$, then there may or may not be a continuous motion $\mathbf{p}(t) = (\mathbf{p}_1(t), \mathbf{p}_2(t), \ldots, \mathbf{p}_N(t))$, with $\mathbf{p}_i(t) \in \mathbb{E}^d$ for all $0 \leq t \leq 1$ and $1 \leq i \leq N$ such that $\mathbf{p}(0) = \mathbf{p}$ and $\mathbf{p}(1) = \mathbf{q}$, and $\|\mathbf{p}_i(t) - \mathbf{p}_j(t)\|$ is monotone decreasing for all $1 \leq i < j \leq N$. When there is such a motion, we say that $\mathbf{q}$ is a *continuous contraction* of $\mathbf{p}$. Finally, let $\mathbf{B}^d[\mathbf{p}_i, r_i]$ denote the (closed) d-dimensional ball centered at $\mathbf{p}_i$ with radius r_i in $\mathbb{E}^d$ and let $\mathrm{vol}_d(\ldots)$ represent the d-dimensional volume (Lebesgue measure) in $\mathbb{E}^d$. In 1954 Poulsen [216] and in 1955 Kneser [183] independently conjectured the following for the case when $r_1 = \cdots = r_N$.

Conjecture 5.1.1 *If* $\mathbf{q} = (\mathbf{q}_1, \mathbf{q}_2, \ldots, \mathbf{q}_N)$ *is a contraction of* $\mathbf{p} = (\mathbf{p}_1, \mathbf{p}_2, \ldots, \mathbf{p}_N)$ *in* $\mathbb{E}^d$, *then*

$$\mathrm{vol}_d \left(\bigcup_{i=1}^{N} \mathbf{B}^d[\mathbf{p}_i, r_i] \right) \geq \mathrm{vol}_d \left(\bigcup_{i=1}^{N} \mathbf{B}^d[\mathbf{q}_i, r_i] \right).$$

Conjecture 5.1.2 *If* $\mathbf{q} = (\mathbf{q}_1, \mathbf{q}_2, \ldots, \mathbf{q}_N)$ *is a contraction of* $\mathbf{p} = (\mathbf{p}_1, \mathbf{p}_2, \ldots, \mathbf{p}_N)$ *in* $\mathbb{E}^d$, *then*

$$\mathrm{vol}_d \left(\bigcap_{i=1}^{N} \mathbf{B}^d[\mathbf{p}_i, r_i] \right) \leq \mathrm{vol}_d \left(\bigcap_{i=1}^{N} \mathbf{B}^d[\mathbf{q}_i, r_i] \right).$$

K. Bezdek, *Classical Topics in Discrete Geometry*, CMS Books in Mathematics, 47
DOI 10.1007/978-1-4419-0600-7_5, © Springer Science+Business Media, LLC 2010

Actually, Kneser seems to be the one who has generated a great deal of interest in the above conjectures also via private letters written to a number of mathematicians. For more details on this see, for example, [181].

5.2 The Kneser–Poulsen Conjecture for Continuous Contractions

For a given point configuration $\mathbf{p} = (\mathbf{p}_1, \mathbf{p}_2, \ldots, \mathbf{p}_N)$ in $\mathbb{E}^d$ and radii $r_1, r_2, \ldots, r_N$ consider the following sets,

$$\mathbf{V}_i = \{\mathbf{x} \in \mathbb{E}^d \mid \text{for all } j, \|\mathbf{x} - \mathbf{p}_i\|^2 - r_i^2 \leq \|\mathbf{x} - \mathbf{p}_j\|^2 - r_j^2\},$$

$$\mathbf{V}^i = \{\mathbf{x} \in \mathbb{E}^d \mid \text{for all } j, \|\mathbf{x} - \mathbf{p}_i\|^2 - r_i^2 \geq \|\mathbf{x} - \mathbf{p}_j\|^2 - r_j^2\}.$$

The set $\mathbf{V}_i$ (resp., $\mathbf{V}^i$) is called the *nearest (resp., farthest) point Voronoi cell* of the point $\mathbf{p}_i$. (For a detailed discussion on nearest as well as farthest point Voronoi cells we refer the interested reader to [124] and [230].) We now restrict each of these sets as follows.

$$\mathbf{V}_i(r_i) = \mathbf{V}_i \cap \mathbf{B}^d[\mathbf{p}_i, r_i],$$

$$\mathbf{V}^i(r_i) = \mathbf{V}^i \cap \mathbf{B}^d[\mathbf{p}_i, r_i].$$

We call the set $\mathbf{V}_i(r_i)$ (resp., $\mathbf{V}^i(r_i)$) the *nearest (resp., farthest) point truncated Voronoi cell* of the point $\mathbf{p}_i$. For each $i \neq j$ let $W_{ij} = \mathbf{V}_i \cap \mathbf{V}_j$ and $W^{ij} = \mathbf{V}^i \cap \mathbf{V}^j$. The sets W_{ij} and W^{ij} are the *walls* between the nearest and farthest point Voronoi cells. Finally, it is natural to define the relevant *truncated walls* as follows.

$$W_{ij}(\mathbf{p}_i, r_i) = W_{ij} \cap \mathbf{B}^d[\mathbf{p}_i, r_i]$$
$$= W_{ij}(\mathbf{p}_j, r_j) = W_{ij} \cap \mathbf{B}^d[\mathbf{p}_j, r_j],$$

$$W^{ij}(\mathbf{p}_i, r_i) = W^{ij} \cap \mathbf{B}^d[\mathbf{p}_i, r_i]$$
$$= W^{ij}(\mathbf{p}_j, r_j) = W^{ij} \cap \mathbf{B}^d[\mathbf{p}_j, r_j].$$

The following formula discovered by Csikós [113] proves Conjecture 5.1.1 as well as Conjecture 5.1.2 for continuous contractions in a straighforward way in any dimension. (Actually, the planar case of the Kneser–Poulsen conjecture under continuous contractions has been proved independently in [77], [112], [99], and [26].)

Theorem 5.2.1 *Let $d \geq 2$ and let $\mathbf{p}(t), 0 \leq t \leq 1$ be a smooth motion of a point configuration in $\mathbb{E}^d$ such that for each t, the points of the configuration are pairwise distinct. Then*

$$\frac{d}{dt}\mathrm{vol}_d \left(\bigcup_{i=1}^{N} \mathbf{B}^d[\mathbf{p}_i(t), r_i] \right)$$

$$= \sum_{1 \leq i < j \leq N} \left(\frac{d}{dt}d_{ij}(t) \right) \cdot \mathrm{vol}_{d-1}\left(W_{ij}(\mathbf{p}_i(t), r_i) \right),$$

$$\frac{d}{dt}\mathrm{vol}_d \left(\bigcap_{i=1}^{N} \mathbf{B}^d[\mathbf{p}_i(t), r_i] \right)$$

$$= \sum_{1 \leq i < j \leq N} -\left(\frac{d}{dt}d_{ij}(t) \right) \cdot \mathrm{vol}_{d-1}\left(W^{ij}(\mathbf{p}_i(t), r_i) \right),$$

where $d_{ij}(t) = \|\mathbf{p}_i(t) - \mathbf{p}_i(t)\|$.

On the one hand, Csikós [114] managed to generalize his formula to configurations of balls called flowers which are sets obtained from balls with the help of operations $\cap$ and $\cup$. This work extends to hyperbolic as well as spherical space. On the other hand, Csikós [115] has succeeded in proving a Schläfli-type formula for polytopes with curved faces lying in pseudo-Riemannian Einstein manifolds, which can be used to provide another proof of Conjecture 5.1.1 as well as Conjecture 5.1.2 for continuous contractions (for more details see [115]).

5.3 The Kneser–Poulsen Conjecture in the Plane

In the recent paper [58] the author and Connelly proved Conjecture 5.1.1 as well as Conjecture 5.1.2 in the Euclidean plane. Thus, we have the following theorem.

Theorem 5.3.1 *If $\mathbf{q} = (\mathbf{q}_1, \mathbf{q}_2, \ldots, \mathbf{q}_N)$ is a contraction of $\mathbf{p} = (\mathbf{p}_1, \mathbf{p}_2, \ldots, \mathbf{p}_N)$ in $\mathbb{E}^2$, then*

$$\mathrm{vol}_2 \left(\bigcup_{i=1}^{N} \mathbf{B}^2[\mathbf{p}_i, r_i] \right) \geq \mathrm{vol}_2 \left(\bigcup_{i=1}^{N} \mathbf{B}^2[\mathbf{q}_i, r_i] \right);$$

moreover,

$$\mathrm{vol}_2 \left(\bigcap_{i=1}^{N} \mathbf{B}^2[\mathbf{p}_i, r_i] \right) \leq \mathrm{vol}_2 \left(\bigcap_{i=1}^{N} \mathbf{B}^2[\mathbf{q}_i, r_i] \right).$$

In fact, the paper [58] contains a proof of an extension of the above theorem to flowers as well. In what follows we give an outline of the three-step proof published in [58] by phrasing it through a sequence of theorems each being higher-dimensional. Voronoi cells play an essential role in our proofs of Theorems 5.3.2 and 5.3.3.

Theorem 5.3.2 *Consider N moving closed d-dimensional balls $\mathbf{B}^d[\mathbf{p}_i(t), r_i]$ with $1 \leq i \leq N, 0 \leq t \leq 1$ in $\mathbb{E}^d, d \geq 2$. If $F_i(t)$ is the contribution of the ith ball to the boundary of the union $\bigcup_{i=1}^N \mathbf{B}^d[\mathbf{p}_i(t), r_i]$ (resp., of the intersection $\bigcap_{i=1}^N \mathbf{B}^d[\mathbf{p}_i(t), r_i]$), then*

$$\sum_{1 \leq i \leq N} \frac{1}{r_i} \operatorname{svol}_{d-1}(F_i(t))$$

decreases (resp., increases) in t under any analytic contraction $\mathbf{p}(t)$ of the center points, where $0 \leq t \leq 1$ and $\operatorname{svol}_{d-1}(\dots)$ refers to the relevant $(d-1)$-dimensional surface volume.

Theorem 5.3.3 *Let the centers of the closed d-dimensional balls $\mathbf{B}^d[\mathbf{p}_i, r_i]$, $1 \leq i \leq N$ lie in the $(d-2)$-dimensional affine subspace L of $\mathbb{E}^d, d \geq 3$. If F_i stands for the contribution of the ith ball to the boundary of the union $\bigcup_{i=1}^N \mathbf{B}^d[\mathbf{p}_i, r_i]$ (resp., of the intersection $\bigcap_{i=1}^N \mathbf{B}^d[\mathbf{p}_i, r_i]$), then*

$$\operatorname{vol}_{d-2}\left(\bigcup_{i=1}^N \mathbf{B}^{d-2}[\mathbf{p}_i, r_i]\right) = \frac{1}{2\pi} \sum_{1 \leq i \leq N} \frac{1}{r_i} \operatorname{svol}_{d-1}(F_i)$$

$$\left(resp., \ \operatorname{vol}_{d-2}\left(\bigcap_{i=1}^N \mathbf{B}^{d-2}[\mathbf{p}_i, r_i]\right) = \frac{1}{2\pi} \sum_{1 \leq i \leq N} \frac{1}{r_i} \operatorname{svol}_{d-1}(F_i)\right),$$

where $\mathbf{B}^{d-2}[\mathbf{p}_i, r_i] = \mathbf{B}^d[\mathbf{p}_i, r_i] \cap L, 1 \leq i \leq N$.

Theorem 5.3.4 *If $\mathbf{q} = (\mathbf{q}_1, \mathbf{q}_2, \dots, \mathbf{q}_N)$ is a contraction of $\mathbf{p} = (\mathbf{p}_1, \mathbf{p}_2, \dots, \mathbf{p}_N)$ in $\mathbb{E}^d, d \geq 1$, then there is an analytic contraction $\mathbf{p}(t) = (\mathbf{p}_1(t), \dots, \mathbf{p}_N(t)), 0 \leq t \leq 1$ in $\mathbb{E}^{2d}$ such that $\mathbf{p}(0) = \mathbf{p}$ and $\mathbf{p}(1) = \mathbf{q}$.*

Note that Theorems 5.3.2, 5.3.3, and 5.3.4 imply Theorem 5.3.1 in a straightforward way.

Also, we note that Theorem 5.3.4 (called the Leapfrog Lemma) cannot be improved; namely, it has been shown in [24] that there exist point configurations $\mathbf{q}$ and $\mathbf{p}$ in $\mathbb{E}^d$, actually constructed in the way suggested in [58], such that $\mathbf{q}$ is a contraction of $\mathbf{p}$ in $\mathbb{E}^d$ and there is no continuous contraction from $\mathbf{p}$ to $\mathbf{q}$ in $\mathbb{E}^{2d-1}$.

In order to describe a more complete picture of the status of the Kneser–Poulsen conjecture we mention two additional corollaries obtained from the proof published in [58] and just outlined above. (For more details see [58].)

Theorem 5.3.5 *Let* $\mathbf{p} = (\mathbf{p}_1, \mathbf{p}_2, \ldots, \mathbf{p}_N)$ *and* $\mathbf{q} = (\mathbf{q}_1, \mathbf{q}_2, \ldots, \mathbf{q}_N)$ *be two point configurations in* $\mathbb{E}^d$ *such that* $\mathbf{q}$ *is a piecewise-analytic contraction of* $\mathbf{p}$ *in* $\mathbb{E}^{d+2}$. *Then the conclusions of Conjecture 5.1.1 as well as Conjecture 5.1.2 hold in* $\mathbb{E}^d$.

The following generalizes a result of Gromov in [150], who proved it in the case $N \leq n + 1$.

Theorem 5.3.6 *If* $\mathbf{q} = (\mathbf{q}_1, \mathbf{q}_2, \ldots, \mathbf{q}_N)$ *is an arbitrary contraction of* $\mathbf{p} = (\mathbf{p}_1, \mathbf{p}_2, \ldots, \mathbf{p}_N)$ *in* $\mathbb{E}^d$ *and* $N \leq n + 3$, *then both Conjecture 5.1.1 and Conjecture 5.1.2 hold.*

As a next step it would be natural to investigate the case $N = n + 4$.

5.4 Non-Euclidean Kneser–Poulsen-Type Results

It is somewhat surprising that in spherical space for the specific radius of balls (i.e., spherical caps) one can find a proof of both Conjecture 5.1.1 and Conjecture 5.1.2 in all dimensions. The magic radius is $\frac{\pi}{2}$ and the following theorem describes the desired result in details.

Theorem 5.4.1 *If a finite set of closed d-dimensional balls of radius $\frac{\pi}{2}$ (i.e., of closed hemispheres) in the d-dimensional spherical space $\mathbb{S}^d, d \geq 2$ is rearranged so that the (spherical) distance between each pair of centers does not increase, then the (spherical) d-dimensional volume of the intersection does not decrease and the (spherical) d-dimensional volume of the union does not increase.*

The method of the proof published by the author and Connelly in [61] can be described as follows. First, one can use a leapfrog lemma to move one configuration to the other in an analytic and monotone way, but only in higher dimensions. Then the higher-dimensional balls have their combined volume (their intersections or unions) change monotonically, a fact that one can prove using Schläfli's differential formula. Then one can apply an integral formula to relate the volume of the higher-dimensional object to the volume of the lower-dimensional object, obtaining the volume inequality for the more general discrete motions.

The following statement is a corollary of Theorem 5.4.1, the Euclidean part of which has been proved independently by Alexander [4], Capoyleas, and Pach [98] and Sudakov [234]. For the sake of completeness in what follows, we recall the notion of spherical mean width, which is most likely less known than its widely used Euclidean counterpart. Let $\mathbb{S}^d$ be the d-dimensional unit sphere centered at the origin in $\mathbb{E}^{d+1}$. A spherically convex body is a closed, spherically convex subset of $\mathbb{S}^d$ with interior points and lying in some closed hemisphere, thus, the intersection of $\mathbb{S}^d$ with a $(d+1)$-dimensional closed convex cone of $\mathbb{E}^{d+1}$ different from $\mathbb{E}^{d+1}$. Recall that $\mathrm{Svol}_d(\ldots)$ denotes the spherical Lebesgue measure on $\mathbb{S}^d$, and recall that $(d+1)\omega_{d+1} = \mathrm{Svol}_d(\mathbb{S}^d)$. Moreover, as usual we denote the standard inner product of $\mathbb{E}^{d+1}$ by $\langle \cdot, \cdot \rangle$, and for $\mathbf{u} \in \mathbb{S}^d$ we write $\mathbf{u}^\perp := \{\mathbf{x} \in \mathbb{E}^{d+1} : \langle \mathbf{u}, \mathbf{x} \rangle = 0\}$ for the orthogonal complement of $\mathrm{lin}\{\mathbf{u}\}$. For a spherically convex body $\mathbf{K}$, the polar body is defined by

$$\mathbf{K}^* := \{\mathbf{u} \in \mathbb{S}^d : \langle \mathbf{u}, \mathbf{v} \rangle \leq 0 \text{ for all } \mathbf{v} \in \mathbf{K}\}.$$

It is also spherically convex, but need not have interior points. The number

$$U(\mathbf{K}) := \frac{1}{2} \mathrm{Svol}_d(\{\mathbf{u} \in \mathbb{S}^d : \mathbf{u}^\perp \cap \mathbf{K} \neq \emptyset\})$$

can be considered as the *spherical mean width* of $\mathbf{K}$. Obviously, a vector $\mathbf{u} \in \mathbb{S}^d$ satisfies $\mathbf{u} \in \mathbf{K}^* \cup (-\mathbf{K}^*)$ if and only if $\mathbf{u}^\perp$ does not meet the interior of $\mathbf{K}$, hence

$$(d+1)\omega_{d+1} - 2\mathrm{Svol}_d(\mathbf{K}^*) = 2U(\mathbf{K}). \tag{5.1}$$

Now, (5.1) and Theorem 5.4.1 imply the following theorem in a rather straighforward way.

Theorem 5.4.2 *Let* $\mathbf{p} = (\mathbf{p}_1, \mathbf{p}_2, \ldots, \mathbf{p}_N)$ *be* N *points on a closed hemisphere of* $\mathbb{S}^d, d \geq 2$ *(resp., points in* $\mathbb{E}^d, d \geq 2$), *and let* $\mathbf{q} = (\mathbf{q}_1, \mathbf{q}_2, \ldots, \mathbf{q}_N)$ *be a contraction of* $\mathbf{p}$ *in* $\mathbb{S}^d$ *(resp., in* $\mathbb{E}^d$). *Then the spherical mean width (resp., mean width) of the spherical convex hull (resp., convex hull) of* $\mathbf{q}$ *is less than or equal to the spherical mean width (resp., mean width) of the spherical convex hull (resp., convex hull) of* $\mathbf{p}$.

Before we continue our non-Euclidean discussions it seems natural to mention a Euclidean Kneser–Poulsen-type result supported by Theorem 5.4.2. For that purpose, let $\mathbf{p} = (\mathbf{p}_1, \mathbf{p}_2, \ldots, \mathbf{p}_N)$ be N points in $\mathbb{E}^d, d \geq 2$, and let $\mathbf{q} = (\mathbf{q}_1, \mathbf{q}_2, \ldots, \mathbf{q}_N)$ be an arbitrary contraction of $\mathbf{p}$ in $\mathbb{E}^d$. Now, if $r > 0$ is sufficiently large, then the union of the balls of radius r centered at the points of $\mathbf{q}$ (resp., $\mathbf{p}$) is eventually the same as the outer parallel domain of radius r of the convex hull of $\mathbf{q}$ (resp., $\mathbf{p}$). Then writing out Steiner's formula for the volumes of the outer parallel domains just mentioned with coefficients equal to the proper intrinsic volumes and noting that the first intrinsic volume is equal to the mean width (up to some constant), Theorem 5.4.2 implies that Conjecture 5.1.1 holds for sufficiently large equal radii (provided of course, that the mean width in question is non-zero). A similar argument supports the inequality of Conjecture 5.1.2 to hold for sufficiently large equal radii. Thus we have arrived at the following theorem that was proved regorously by Gorbovickis in [149] (using a different approach).

Theorem 5.4.3 *If* $\mathbf{q} = (\mathbf{q}_1, \mathbf{q}_2, \ldots, \mathbf{q}_N)$ *is a contraction of* $\mathbf{p} = (\mathbf{p}_1, \mathbf{p}_2, \ldots, \mathbf{p}_N)$ *in* $\mathbb{E}^d$, *then there exists* $r_0 > 0$ *such that for any* $r \geq r_0$,

$$\mathrm{vol}_d \left(\bigcup_{i=1}^N \mathbf{B}^d[\mathbf{p}_i, r] \right) \geq \mathrm{vol}_d \left(\bigcup_{i=1}^N \mathbf{B}^d[\mathbf{q}_i, r] \right)$$

$$\left(resp., \ \mathrm{vol}_d \left(\bigcap_{i=1}^{N} \mathbf{B}^d[\mathbf{p}_i, r] \right) \leq \mathrm{vol}_d \left(\bigcap_{i=1}^{N} \mathbf{B}^d[\mathbf{q}_i, r] \right) \right).$$

We note that Theorem 5.4.1 extends to flowers as well; moreover, a positive answer to the following problem would imply that both Conjecture 5.1.1 and Conjecture 5.1.2 hold for circles in $\mathbb{S}^2$ (for more details on this see [61]).

Problem 5.4.4 *Suppose that* $\mathbf{p} = (\mathbf{p}_1, \mathbf{p}_2, \dots, \mathbf{p}_N)$ *and* $\mathbf{q} = (\mathbf{q}_1, \mathbf{q}_2, \dots, \mathbf{q}_N)$ *are two point configurations in* $\mathbb{S}^2$. *Then prove or disprove that there is a monotone piecewise-analytic motion from* $\mathbf{p} = (\mathbf{p}_1, \mathbf{p}_2, \dots, \mathbf{p}_N)$ *to* $\mathbf{q} = (\mathbf{q}_1, \mathbf{q}_2, \dots, \mathbf{q}_N)$ *in* $\mathbb{S}^4$.

Note that in fact, Theorem 5.4.1 states a volume inequality between two spherically convex polytopes satisfying some metric conditions. The following problem searches for a natural analogue of that in the hyperbolic 3-space $\mathbb{H}^3$. In order to state it properly we recall the following. Let A and B be two planes in $\mathbb{H}^3$ and let A^+ (resp., B^+) denote one of the two closed halfspaces bounded by A (resp., B) such that the set $A^+ \cap B^+$ is nonempty. Recall that either A and B intersect or A is parallel to B or A and B have a line perpendicular to both of them. Now, "the dihedral angle $A^+ \cap B^+$" means not only the set in question, but also refers to the standard angular measure of the corresponding angle between A and B in the first case, it refers to 0 in the second case, and finally, in the third case it refers to the negative of the hyperbolic distance between A and B.

Problem 5.4.5 *Let* $\mathbf{P}$ *and* $\mathbf{Q}$ *be compact convex polyhedra of* $\mathbb{H}^3$ *with* $\mathbf{P}$ *(resp.,* $\mathbf{Q}$*) being the intersection of the closed halfspaces* $H_{P,1}^+, H_{P,2}^+, \dots, H_{P,N}^+$ *(resp.,* $H_{Q,1}^+, H_{Q,2}^+, \dots, H_{Q,N}^+$*). Assume that the dihedral angle* $H_{Q,i}^+ \cap H_{Q,j}^+$ *(containing* $\mathbf{Q}$*) is at least as large as the corresponding dihedral angle* $H_{P,i}^+ \cap H_{P,j}^+$ *(containing* $\mathbf{P}$*) for all* $1 \leq i < j \leq N$. *Then prove or disprove that the volume of* $\mathbf{P}$ *is at least as large as the volume of* $\mathbf{Q}$.

Using Andreev's version [6], [7] of the Koebe–Andreev–Thurston theorem and Schläfli's differential formula the author [64] proved the following partial analogue of Theorem 5.4.1 in $\mathbb{H}^3$.

Theorem 5.4.6 *Let* $\mathbf{P}$ *and* $\mathbf{Q}$ *be nonobtuse-angled compact convex polyhedra of the same simple combinatorial type in* $\mathbb{H}^3$. *If each inner dihedral angle of* $\mathbf{Q}$ *is at least as large as the corresponding inner dihedral angle of* $\mathbf{P}$, *then the volume of* $\mathbf{P}$ *is at least as large as the volume of* $\mathbf{Q}$.

5.5 Alexander's Conjecture

It seems that in the Euclidean plane, for the case of the intersection of congruent disks, one can sharpen the results proved by the author and Connelly [58]. Namely, Alexander [4] conjectures the following.

Conjecture 5.5.1 *Under arbitrary contraction of the center points of finitely many congruent disks in the Euclidean plane, the perimeter of the intersection of the disks cannot decrease.*

The analogous question for the union of congruent disks has a negative answer, as was observed by Habicht and Kneser long ago (for details see [58]). In [68] some supporting evidence for the above conjecture of Alexander has been collected; in particular, the following theorem was proved.

Theorem 5.5.2 *Alexander's conjecture holds for continuous contractions of the center points and it holds up to 4 congruent disks under arbitrary contractions of the center points.*

We note that Alexander's conjecture does not hold for incongruent disks (even under continuous contractions of their center points) as shown in [68]. Finally we remark that if Alexander's conjecture were true, then it would be a rare instance of an asymmetry between intersections and unions for Kneser–Poulsen-type questions.

5.6 Densest Finite Sphere Packings

Let $\mathbf{B}^d$ denote the closed d-dimensional unit ball centered at the origin $\mathbf{o}$ of $\mathbb{E}^d, d \geq 2$ and let $\mathcal{P} := \{\mathbf{c}_1 + \mathbf{B}^d, \mathbf{c}_2 + \mathbf{B}^d, \ldots, \mathbf{c}_n + \mathbf{B}^d\}$ be a packing of n unit balls with centers $\mathbf{c}_1, \mathbf{c}_2, \ldots, \mathbf{c}_n$ in $\mathbb{E}^d$. We say that $\mathcal{P}$ is a *densest packing* among all packings of n unit balls in $\mathbb{E}^d$ if there exists a parameter $r > 1$ with the property that

$$\delta(\mathcal{P}) := \frac{n\mathrm{vol}_d(\mathbf{B}^d)}{\mathrm{vol}_d\left(\bigcup_{i=1}^{n} \mathbf{c}_i + r\mathbf{B}^d\right)} = \frac{n\omega_d}{\mathrm{vol}_d\left(\bigcup_{i=1}^{n} \mathbf{c}_i + r\mathbf{B}^d\right)}$$

$$= \max\left\{\frac{n\omega_d}{\mathrm{vol}_d\left(\bigcup_{i=1}^{n} \mathbf{x}_i + r\mathbf{B}^d\right)} \mid \|\mathbf{x}_j - \mathbf{x}_k\| \geq 2 \text{ for all } 1 \leq j < k \leq n\right\};$$

that is,

$$\mathrm{vol}_d\left(\bigcup_{i=1}^{n} \mathbf{c}_i + r\mathbf{B}^d\right) = \min_{\|\mathbf{x}_j - \mathbf{x}_k\| \geq 2 \text{ for all } 1 \leq j < k \leq n} \left\{\mathrm{vol}_d\left(\bigcup_{i=1}^{n} \mathbf{x}_i + r\mathbf{B}^d\right)\right\}. \quad (5.2)$$

The definition (5.2) is rather natural from the point of the Kneser–Poulsen Conjecture and it seems to lead to a new definition of densest finite sphere packings. The closest related notion is the definition of parametric density, introduced by Wills in [247] (see also [28]), where the union of balls is replaced by the convex hull of the union of balls thereby replacing our concave container by a convex one.

First, let us investigate (5.2) in $\mathbb{E}^2$. If (5.2) holds with parameter r satisfying $1 < r \leq \frac{2}{\sqrt{3}} = 1.1547\ldots$, then it is easy to see that $\mathcal{P}$ must be a packing with the largest number of touching pairs among all packings of n unit disks, and therefore according to the well-known result of Harborth [166], $\mathcal{P}$ must be a subset of the densest infinite hexagonal packing of unit disks in $\mathbb{E}^2$. If (5.2) holds with parameter r satisfying $\frac{2}{\sqrt{3}} < r$, then the Hajós Lemma (see, for example, [200]) easily implies that $\delta(\mathcal{P}) < \frac{\pi}{\sqrt{12}}$. This inequality, for any fixed $\frac{2}{\sqrt{3}} < r$, is asymptotically best possible (with respect to n). However, the following remains a challenging open question.

Problem 5.6.1 *Assume that $\mathcal{P}$ is a densest packing of n unit disks in $\mathbb{E}^2$ with parameter $\frac{2}{\sqrt{3}} < r$ in (5.2). Prove or disprove that $\mathcal{P}$ is a subset of the densest infinite hexagonal packing of unit disks in $\mathbb{E}^2$.*

Next, let us take a closer look of (5.2) in $\mathbb{E}^3$. If (5.2) holds with parameter r satisfying $2 \leq r$, then Theorem 2.4.3 and Theorem 1.4.1 imply in a straightforward way that $\delta(\mathcal{P}) \leq \frac{\pi}{\sqrt{18}}$. Not surprisingly, this inequality, for any fixed $2 \leq r$, is asymptotically best possible (with respect to n). Moreover, if (5.2) holds with parameter r satisfying $\sqrt{\frac{3}{2}} = 1.2247\cdots \leq r < 2$, then Theorem 1.4.6 implies that $\delta(\mathcal{P}) \leq \sigma_3 = 0.7796\ldots$. Last but not least, if (5.2) holds with parameter r satisfying $1 < r < \frac{2}{\sqrt{3}} = 1.1547\ldots$, then it is easy to see that $\mathcal{P}$ must be a packing with the largest number $C(n)$ of touching pairs among all packings of n unit balls in $\mathbb{E}^3$. For some exact values as well as estimates on $C(n)$ see Theorem 1.3.5 and the discussion there. The following problem might generate further progress on the problem at hand. For natural reasons we call it the *Truncated Dodecahedral Conjecture.*

Conjecture 5.6.2 *Let $\mathcal{F}$ be an arbitrary (finite or infinite) family of non-overlapping unit balls in $\mathbb{E}^3$ with the unit ball $\mathbf{B}$ centered at the origin $\mathbf{o}$ of $\mathbb{E}^3$ belonging to $\mathcal{F}$. Let $\mathbf{P}$ stand for the Voronoi cell of the packing $\mathcal{F}$ assigned to $\mathbf{B}$ and let $\mathbf{Q}$ denote a regular dodecahedron circumscribed $\mathbf{B}$ having circumradius $\sqrt{3}\tan\frac{\pi}{5} = 1.2584\ldots$. If r is any parameter with $\frac{2}{\sqrt{3}} < r \leq \sqrt{3}\tan\frac{\pi}{5}$, then*

$$\mathrm{vol}_3(\mathbf{P} \cap r\mathbf{B}) \geq \mathrm{vol}_3(\mathbf{Q} \cap r\mathbf{B}) \ .$$

We note that obviously the inequality of Conjecture 5.6.2 holds for any parameter with $1 < r \leq \frac{2}{\sqrt{3}}$. Moreover, for the sake of completeness we mention that the special case, when $r = \sqrt{3}\tan\frac{\pi}{5}$ in Conjecture 5.6.2, had already been conjectured by L. Fejes Tóth in [135], and it is still open, although the closely related (but weaker) Dodecahedral Conjecture has been recently proved by Hales and McLaughlin [164], [165].

Finally, we take a look at (5.2) in $\mathbb{E}^d, d \geq 4$. On the one hand, if (5.2) holds with parameter r satisfying $2 \leq r$, then Theorem 2.4.3 implies the estimate $\delta(\mathcal{P}) \leq \delta(\mathbf{B}^d)$. On the other hand, if (5.2) holds with parameter r

satisfying $\sqrt{\frac{2d}{d+1}} \leq r < 2$, then Theorem 1.4.6 implies that $\delta(\mathcal{P}) \leq \sigma_d$. In fact, Theorem 1.4.8 improves that inequality to $\delta(\mathcal{P}) \leq \hat{\sigma}_d \ (< \sigma_d)$ for all $d \geq 8$. Last but not least, if (5.2) holds with parameter r satisfying $1 < r < \frac{2}{\sqrt{3}}$, then it is easy to see that $\mathcal{P}$ must be a packing with the largest number of touching pairs (called the contact number of $\mathcal{P}$) among all packings of n unit balls in $\mathbb{E}^d$. Theorem 2.4.2 gives estimates on the contact number of $\mathcal{P}$.

6

Ball-Polyhedra as Intersections of Congruent Balls

6.1 Disk-Polygons and Ball-Polyhedra

The previous sections indicate a good deal of geometry on unions and intersections of balls that is worthwhile studying. In particular, when we restrict our attention to intersections of balls the underlying convexity suggests a broad spectrum of new analytic and combinatorial results. To make the setup ideal for discrete geometry from now on we look at intersections of finitely many congruent closed d-dimensional balls with non-empty interior in $\mathbb{E}^d$. In fact, one may assume that the congruent d-dimensional balls in question are of unit radius; that is, they are unit balls of $\mathbb{E}^d$. Also, it is natural to assume that removing any of the unit balls defining the intersection in question yields the intersection of the remaining unit balls becoming a larger set. If $d = 2$, then we call the sets in question *disk-polygons* and for $d \geq 3$ they are called *ball-polyhedra*. This definition along with some basic properties of ball-polyhedra (resp., disk-polygons) were introduced by the author in a sequence of talks at the University of Calgary in the fall of 2004. Based on that, the paper [69] written by the author, Lángi, Naszódi, and Papez systematically extended those investigations to get a better understanding of the geometry of ball-polyhedra (resp., disk-polygons) by proving a number of theorems, which one can regard as the analogues of the classical theorems on convex polytopes.

6.2 Shortest Billiard Trajectories in Disk-Polygons

Billiards have been around for quite some time in mathematics and have generated a great deal of research. (See, for example, the recent elegant book [237] of Tabachnikov.) For our purposes it seems natural to define billiard trajectories in the following way. This introduces a larger class of polygonal paths for billiard trajectories than the traditional definition widely used in the literature (see [237]). Let $\mathbf{C}$ be an arbitrary convex body in the d-dimensional Euclidean space $\mathbb{E}^d$, $d \geq 2$ that is a compact convex set with non-empty interior in $\mathbb{E}^d$. Then we say that the closed polygonal path $\mathbf{P}$ (possibly with self-intersections) is a *generalized billiard trajectory* of $\mathbf{C}$ if all the vertices of $\mathbf{P}$ lie on the boundary of $\mathbf{C}$ and if each of the (inner) angle bisectors of $\mathbf{P}$ between two consecutive sides of $\mathbf{P}$ is perpendicular to a supporting hyperplane of $\mathbf{C}$ passing through the corresponding vertex of $\mathbf{P}$. If $\mathbf{P}$ has n sides, then we say that $\mathbf{P}$ is an *n-periodic generalized billiard trajectory* in $\mathbf{C}$. Note that our definition of generalized billiard trajectories coincides with the traditional definition of billiard trajectories whenever the billiard table has no corner points. It seems that the paper [42] was among the first suggesting a detailed study of generalized billiard trajectories in convex domains. For some analogue higher-dimensional investigations we refer the interested reader to the recent paper [146]. Generalized billiard trajectories have the following fundamental property proved by the author and D. Bezdek in the very recent paper [38].

Theorem 6.2.1 *Let $\mathbf{C}$ be a convex body in $\mathbb{E}^d$, $d \geq 2$. Then $\mathbf{C}$ possesses at least one shortest generalized billiard trajectory; moreover, any of the shortest generalized billiard trajectories in $\mathbf{C}$ is of period at most $d + 1$.*

For the sake of completeness we mention that according to the main result of [25] any d-dimensional billiard table with a smooth boundary, but not necessarily convex, has a k-periodic billiard trajectory with $k \leq d + 1$, which is a closely related result.

It was observed in [38] that for the following special family of disk-polygons one can improve the estimate on periods in Theorem 6.2.1. If $\mathbf{D}$ is a disk-polygon in $\mathbb{E}^2$ having the property that the pairwise distances between the centers of its generating unit disks are at most 1, then we say that $\mathbf{D}$ is a *fat disk-polygon*.

Theorem 6.2.2 *Let $\mathbf{D}$ be a fat disk-polygon in $\mathbb{E}^2$. Then any of the shortest generalized billiard trajectories in $\mathbf{D}$ is a 2-periodic one.*

In the recent paper [146], the following fundamental question is studied that was raised by Zelditch in [249] motivated by applications to inverse spectral problems. In which convex bodies are the shortest periodic billiard trajectories of period 2? It is proved in [146] that any convex body whose inscribed ball touches the boundary of the given convex body at two diametrically opposite points has that property. Theorem 6.2.2 shows that the family of fat disk-polygons possesses the same property as well.

According to Birkhoff's well-known theorem ([237]) if $\mathbf{B}$ is a strictly convex billiard table with smooth boundary (i.e., if the boundary of $\mathbf{B}$ is a simple, closed, smooth, and strictly convex curve) in $\mathbb{E}^2$, then for every positive integer $N > 1$ there exist (at least two) N-periodic billiard trajectories in $\mathbf{B}$. (In fact, here the rotation number of the billiard trajectory in question can be preassigned as well. Also, it is well known that neither the convexity nor smoothness can be removed from the assumptions in order to have the same conclusion. Last but not least, for a higher-dimensional analogue of Birkhoff's theorem we refer the interested reader to [128].) Billiard tables suitable for Birkhoff's theorem can be easily constructed from disk-polygons as follows. Take a disk-polygon $\mathbf{D}$ in $\mathbb{E}^2$. Then choose a positive ϵ not larger than the inradius of $\mathbf{D}$ (which is the radius of the largest circular disk contained in $\mathbf{D}$) and take the union of all circular disks of radius ϵ that lie in $\mathbf{D}$. We call the set obtained in this way the ϵ-rounded disk-polygon of $\mathbf{D}$ and denote it by $\mathbf{D}(\epsilon)$. The following theorem proved in [38] also shows the complexity of the problem of Zelditch [249] on characterizing convex domains whose shortest periodic billiard trajectories are of period 2.

Theorem 6.2.3 *Let $\mathbf{D}$ be a fat disk-polygon in $\mathbb{E}^2$. Then any of the shortest (generalized) billiard trajectories in the ϵ-rounded disk-polygon $\mathbf{D}(\epsilon)$ is a 2-periodic one for all $\epsilon > 0$ being sufficiently small.*

Actually, we believe ([38]) that the following even stronger statement holds.

Conjecture 6.2.4 *Let $\mathbf{D}$ be a fat disk-polygon in $\mathbb{E}^2$. Then any of the shortest (generalized) billiard trajectories in the ϵ-rounded disk-polygon $\mathbf{D}(\epsilon)$ is a 2-periodic one for all ϵ being at most as large as the inradius of $\mathbf{D}$.*

Finally we mention the following result that can be obtained as an immediate corollary of Theorem 6.2.2. This might be of independent interest, in particular because it generalizes the result proved in [42] that any closed curve of length at most 1 can be covered by a translate of any convex domain of constant width $\frac{1}{2}$ in the Euclidean plane. As usual if $\mathbf{C}$ is a convex domain of the Euclidean plane, then let $w(\mathbf{C})$ denote the minimal width of $\mathbf{C}$ (i.e., the smallest distance between two parallel supporting lines of $\mathbf{C}$).

Corollary 6.2.5 *Let $\mathbf{D}$ be a fat disk-polygon in $\mathbb{E}^2$. Then any closed curve of length at most $2w(\mathbf{D})$ in $\mathbb{E}^2$ can be covered by a translate of $\mathbf{D}$.*

It would be interesting to find the higher-dimensional analogues of the 2-dimensional results just mentioned in this section. In particular, the following question does not seem to be an easy one.

Problem 6.2.6 *Let $\mathbf{P}$ be a ball-polyhedron in $\mathbb{E}^d, d \geq 3$ with the property that the pairwise distances between the centers of its generating unit balls are at most 1. Then prove or disprove that any of the shortest generalized billiard trajectories in $\mathbf{P}$ is a 2-periodic one.*

6.3 Blaschke–Lebesgue-Type Theorems for Disk-Polygons

The classical isoperimetric inequality combined with Barbier's theorem (stating that the perimeter of any convex domain of constant width w is equal to

πw) implies that the largest area of convex domains of constant width w is the circular disk of diameter w, having the area of $\frac{\pi}{4}w^2$ (for more details see, for example, [79]). On the other hand, the well-known Blaschke-Lebesgue theorem states that among all convex domains of constant width w, the Reuleaux triangle of width w has the smallest area, namely $\frac{1}{2}(\pi - \sqrt{3})w^2$. Blaschke [76] and Lebesgue [188] were the first to show this and the succeeding decades have seen other works published on different proofs of that theorem. For a most recent new proof, and for a survey on the state of the art of different proofs of the Blaschke–Lebesgue theorem, see the elegant paper of Harrell [167]. The main goal of this section is to extend the Blaschke–Lebesgue Theorem for disk-polygons.

The disk-polygon $\mathbf{D}$ is called a disk-polygon with *center parameter* t, $0 < t < \sqrt{3} = 1.732\ldots$, if the distance between any two centers of the generating unit disks of $\mathbf{D}$ is at most t. Let $\mathcal{F}(t)$ denote the family of all disk-polygons with center parameter t. Let $\mathbf{\Delta}(t)$ denote the regular disk-triangle whose three generating unit disks are centered at the vertices of a regular triangle of side length t, $1 \leq t < \sqrt{3} = 1.732\ldots$. Recall that the *inradius* $r(\mathbf{C})$ of a convex domain $\mathbf{C}$ in $\mathbb{E}^2$ is the radius of the largest circular disk lying in $\mathbf{C}$ (simply called the *incircle* of $\mathbf{C}$). The following formulas give the inradius $r(\mathbf{\Delta}(t))$, the minimal width $w(\mathbf{\Delta}(t))$, the area $a(\mathbf{\Delta}(t))$ and the perimeter $p(\mathbf{\Delta}(t))$ of $\mathbf{\Delta}(t)$ for all $1 \leq t < \sqrt{3}$:

$$r(\mathbf{\Delta}(t)) = 1 - \frac{1}{3}\sqrt{3}t;$$

$$w(\mathbf{\Delta}(t)) = 1 - \frac{1}{2}\sqrt{4 + 2t^2 - 2\sqrt{3}t\sqrt{4 - t^2}};$$

$$a(\mathbf{\Delta}(t)) = \frac{3}{2}\arccos t + \frac{1}{4}\sqrt{3}t^2 - \frac{3}{4}t\sqrt{4 - t^2} - \frac{1}{2}\pi;$$

$$p(\mathbf{\Delta}(t)) = 2\pi - 6\arcsin\frac{t}{2}.$$

The following theorem has been proved by M. Bezdek [75].

Theorem 6.3.1 *Let* $\mathbf{D} \in \mathcal{F}(t)$ *be an arbitrary disk-polygon with center parameter* t, $1 \leq t < \sqrt{3}$. *Then* $r(\mathbf{D}) \geq r(\mathbf{\Delta}(t))$ *and* $w(\mathbf{D}) \geq w(\mathbf{\Delta}(t))$. *Moreover, the area of* $\mathbf{D}$ *is at least as large as the area of* $\mathbf{\Delta}(t)$; *that is,*

$$a(\mathbf{D}) \geq a(\mathbf{\Delta}(t))$$

with equality if and only if $\mathbf{D} = \mathbf{\Delta}(t)$.

For $t = 1$ the above area inequality and the well-known fact (see, e.g., [79]) that the family of Reuleaux polygons of width 1 is a dense subset of the family of convex domains of constant width 1, imply the Blaschke–Lebesgue theorem in a straightforward way. In connection with Theorem 6.3.1 we propose to investigate the following related problem, in particular, because an affirmative answer to that question would imply the area inequality of Theorem 6.3.1.

Problem 6.3.2 *Let* $\mathbf{D} \in \mathcal{F}(t)$ *be an arbitrary disk-polygon with center parameter* $t, 1 \leq t < \sqrt{3}$. *Prove or disprove that the perimeter of* $\mathbf{D}$ *is at least as large as the perimeter of* $\mathbf{\Delta}(t)$; *that is,*

$$p(\mathbf{D}) \geq p(\mathbf{\Delta}(t)).$$

Let $\mathbf{C} \subset \mathbb{E}^2$ be a convex domain and let $\rho > 0$ be given. Then, the *outer parallel domain* $\mathbf{C}_\rho$ of radius ρ of $\mathbf{C}$ is the union of all (closed) circular disks of radii ρ, whose centers belong to $\mathbf{C}$. Recall that $a(\mathbf{C}_\rho) = a(\mathbf{C}) + p(\mathbf{C})\rho + \pi\rho^2$. Let $0 < t < 1$ be given and let $\mathbf{R}(t)_{1-t}$ denote the outer parallel domain of radius $1 - t$ of a Reuleaux triangle $\mathbf{R}(t)$ of width t. Note that $\mathbf{R}(t)_{1-t}$ is a convex domain of constant width $2 - t$ and so, Barbier's theorem ([79]) implies that its perimeter is equal to $p(\mathbf{R}(t)_{1-t}) = \pi(2 - t)$; moreover, it is not hard to check that its area is equal to $a(\mathbf{R}(t)_{1-t}) = \frac{1}{2}(\pi - \sqrt{3})t^2 - \pi t + \pi$. The following theorem is a natural counterpart of Theorem 6.3.1. Also, we note that it is equivalent to the 2-dimensional case of Theorem 4.5.8.

Theorem 6.3.3 *Let* $\mathbf{D} \in \mathcal{F}(t)$ *be an arbitrary disk-polygon with center parameter* $t, 0 < t < 1$. *Then, the area of* $\mathbf{D}$ *is strictly larger than the area of* $\mathbf{R}(t)_{1-t}$; *that is,*

$$a(\mathbf{D}) > a(\mathbf{R}(t)_{1-t}).$$

6.4 On the Steinitz Problem for Ball-Polyhedra

One can represent the boundary of a ball-polyhedron in $\mathbb{E}^3$ as the union of *vertices, edges*, and *faces* defined in a rather natural way as follows. A boundary point is called a *vertex* if it belongs to at least three of the closed unit balls defining the ball-polyhedron. A *face* of the ball-polyhedron is the intersection of one of the generating closed unit balls with the boundary of the ball-polyhedron. Finally, if the intersection of two faces is non-empty, then it is the union of (possibly degenerate) circular arcs. The non-degenerate arcs are called *edges* of the ball-polyhedron. Obviously, if a ball-polyhedron in $\mathbb{E}^3$ is generated by at least three unit balls, then it possesses vertices, edges, and faces. Finally, a ball-polyhedron is called a *standard ball-polyhedron* if its vertices, edges, and faces (together with the empty set and the ball-polyhedron itself) form an algebraic lattice with respect to containment. We note that not every ball-polyhedron of $\mathbb{E}^3$ is a standard one, a fact that is somewhat surprising and is responsible for some of the difficulties arising in studying ball-polyhedra in general (for more details see [66] as well as [69]).

For us a *graph* is always a non-oriented one that has finitely many vertices and edges. Also, recall that a graph is 3-*connected* if it has at least four vertices and deleting any two vertices yields a connected graph. Moreover, a graph is called *simple* if it contains no loops (edges with identical endpoints) and no parallel edges (edges with the same two endpoints). Finally, a graph is *planar* if it can be drawn in the Euclidean plane without crossing edges. Now, recall that according to the well-known theorem of Steinitz a graph is the edge-graph of some convex polyhedron in $\mathbb{E}^3$ if, and only if, it is simple, planar, and 3-connected. As a partial analogue of Steinitz's theorem for ball-polyhedra the following theorem is proved in [69].

Theorem 6.4.1 *The edge-graph of any standard ball-polyhedron in $\mathbb{E}^3$ is a simple, planar, and 3-connected graph.*

Based on that it would be natural to look for an answer to the following question raised in [69].

Problem 6.4.2 *Prove or disprove that every simple, planar, and 3-connected graph is the edge-graph of some standard ball-polyhedron in $\mathbb{E}^3$.*

6.5 On Global Rigidity of Ball-Polyhedra

One of the best known results in the geometry of convex polyhedra is Cauchy's rigidity theorem: If two convex polyhedra $\mathbf{P}$ and $\mathbf{Q}$ in $\mathbb{E}^3$ are combinatorially equivalent with the corresponding faces being congruent, then the angles between the corresponding pairs of adjacent faces are also equal and thus, $\mathbf{P}$ is congruent to $\mathbf{Q}$. Putting it somewhat differently the combinatorics of an arbitrary convex polyhedron and its face angles completely determine its inner dihedral angles. For more details on Cauchy's rigidity theorem and on its extensions we refer the interested reader to [106].

In this section we look for analogues of Cauchy's rigidity theorem for ball-polyhedra. In order to reach this goal in a short way it seems useful to recall the following terminology from [66]. To each edge of a ball-polyhedron in $\mathbb{E}^3$ we can assign an *inner dihedral angle*. Namely, take any point $\mathbf{p}$ in the relative interior of the edge and take the two unit balls that contain the two faces of the ball-polyhedron meeting along that edge. Now, the inner dihedral angle along this edge is the angular measure of the intersection of the two half-spaces supporting the two unit balls at $\mathbf{p}$. The angle in question is obviously independent of the choice of $\mathbf{p}$. Finally, at each vertex of a face of a ball-polyhedron there is a *face angle* formed by the two edges meeting at the given vertex (which is, in fact, the angle between the two tangent halflines of the two edges meeting at the given vertex). We say that the standard ball-polyhedron $\mathbf{P}$ in $\mathbb{E}^3$ is *globally rigid with respect to its face angles* (resp., *its inner dihedral angles*) if the following holds. If $\mathbf{Q}$ is another standard ball-polyhedron in $\mathbb{E}^3$ whose face lattice is isomorphic to that of $\mathbf{P}$ and whose face angles (resp., inner dihedral angles) are equal to the corresponding face angles (resp. inner dihedral angles) of $\mathbf{P}$, then $\mathbf{Q}$ is congruent to $\mathbf{P}$. A ball-polyhedron of $\mathbb{E}^3$ is called *triangulated* if all its faces are bounded by three edges. It is easy to see that any triangulated ball-polyhedron is, in fact, a standard one.

The claims (*i*) and (*ii*) of Theorem 6.5.1 have been proved in [66]. Claim (*iii*) of Theorem 6.5.1 is based on a special class of standard polyhedra defined as follows. We say that **P** is a *normal ball-polyhedron* if **P** is a standard ball-polyhedron in $\mathbb{E}^3$ with the property that the vertices of the underlying farthest point Voronoi tiling of the center points of the generating unit balls of **P** all belong to the interior of **P**. (Actually, this condition is equivalent to the following one: the distance between any center point of the generating unit balls of **P** and any of the vertices of the farthest point Voronoi cell assigned to the center in question is strictly less than one.) Now, recall that the farthest point Voronoi tiling just mentioned gives rise to the relevant Delaunay tiling of the convex hull **P'** of the centers of the generating unit balls of **P**. This induces a duality between the face lattices of the ball-polyhedron **P** and of the convex polyhedron **P'**. Thus, it is not hard to see that claim (*i*) of Theorem 6.5.1 and Cauchy's rigidity theorem applied to **P'** imply statement (*iii*) of Theorem 6.5.1 on **P**. Thus, we have the following analogues of Cauchy's rigidity theorem for ball-polyhedra.

Theorem 6.5.1
(*i*) *The face lattice and the face angles determine the inner dihedral angles of any standard ball-polyhedron in $\mathbb{E}^3$.*
(*ii*) *Let* **P** *be a triangulated ball-polyhedron in $\mathbb{E}^3$. Then* **P** *is globally rigid with respect to its face angles.*
(*iii*) *Let* **P** *be a normal ball-polyhedron in $\mathbb{E}^3$. Then* **P** *is globally rigid with respect to its face angles.*

Deciding whether all standard ball-polyhedra of $\mathbb{E}^3$ are globally rigid with respect to their face angles remains a challenging open problem. Finally, Theorem 6.5.1 raises the following dual question.

Problem 6.5.2 *Prove or disprove that the face lattice and the inner dihedral angles determine the face angles of any standard ball-polyhedron in $\mathbb{E}^3$.*

We mention that one can regard the above problem as an extension of the (still unresolved) conjecture of Stoker [233] according to which for convex polyhedra the face lattice and the inner dihedral angles determine the face angles.

6.6 Separation and Support for Spindle Convex Sets

The following theorem of Kirchberger is well known (see, e.g., [22]). If A and B are finite (resp., compact) sets in $\mathbb{E}^d$ with the property that for any set $T \subset A \cup B$ of cardinality at most $d + 2$ (i.e., with card $T \le d + 2$) the two sets $A \cap T$ and $B \cap T$ can be strictly separated by a hyperplane, then A

and B can be strictly separated by a hyperplane. It is shown in [69] that no similar statement holds for separation by unit spheres. However, [69] proves the following analogue of Kirchberger's theorem for separation by spheres of radius at most one. For this purpose it is convenient to denote the $(d-1)$-dimensional sphere of $\mathbb{E}^d$ centered at the point $\mathbf{c}$ and having radius r with $S^{d-1}(\mathbf{c}, r)$ and say that the sets $A \subset \mathbb{E}^d$ and $B \subset \mathbb{E}^d$ are *strictly separated* by $S^{d-1}(\mathbf{c}, r)$ if both A and B are disjoint from $S^{d-1}(\mathbf{c}, r)$ and one of them lies in the interior and the other one in the exterior of $S^{d-1}(\mathbf{c}, r)$. Also, we denote by $\mathbf{B}^d(\mathbf{c}, r)$ (resp., $\mathbf{B}^d[\mathbf{c}, r]$) the *open* (resp., *closed*) ball of radius r centered at the point $\mathbf{c}$ in $\mathbb{E}^d$. Thus, we say that the sets A and B are *separated* by $S^{d-1}(\mathbf{c}, r)$ in $\mathbb{E}^d$ if either $A \subset \mathbf{B}^d[\mathbf{c}, r]$ and $B \subset \mathbb{E}^d \setminus \mathbf{B}^d(\mathbf{c}, r)$, or $B \subset \mathbf{B}^d[\mathbf{c}, r]$ and $A \subset \mathbb{E}^d \setminus \mathbf{B}^d(\mathbf{c}, r)$.

Theorem 6.6.1 *Let $A, B \subset \mathbb{E}^d$ be finite sets. Then A and B can be strictly separated by a sphere $S^{d-1}(\mathbf{c}, r)$ with $r \leq 1$ such that $A \subset \mathbf{B}^d(\mathbf{c}, r)$ if and only if the following holds. For every $T \subset A \cup B$ with* card $T \leq d+2$, $T \cap A$ *and $T \cap B$ can be strictly separated by a sphere $S^{d-1}(\mathbf{c}_T, r_T)$ with $r_T \leq 1$ such that $T \cap A \subset \mathbf{B}^d(\mathbf{c}_T, r_T)$.*

The following interesting question remains open.

Problem 6.6.2 *Prove or disprove that Theorem 6.6.1 extends to compact sets.*

It is natural to proceed with some basic support and separation properties of convex sets of special kind that include ball-polyhedra. For this purpose let us recall the following definition from [69]. Let $\mathbf{a}$ and $\mathbf{b}$ be two points in $\mathbb{E}^d$. If $\|\mathbf{a} - \mathbf{b}\| < 2$, then the *(closed) spindle* of $\mathbf{a}$ and $\mathbf{b}$, denoted by $[\mathbf{a}, \mathbf{b}]_s$, is defined as the union of circular arcs with endpoints $\mathbf{a}$ and $\mathbf{b}$ that are of radii at least one and are shorter than a semicircle. If $\|\mathbf{a} - \mathbf{b}\| = 2$, then $[\mathbf{a}, \mathbf{b}]_s := \mathbf{B}^d(\frac{\mathbf{a}+\mathbf{b}}{2}, 1)$. If $\|\mathbf{a} - \mathbf{b}\| > 2$, then we define $[\mathbf{a}, \mathbf{b}]_s$ to be $\mathbb{E}^d$. Next, a set $\mathbf{C} \subset \mathbb{E}^d$ is called *spindle convex* if, for any pair of points $\mathbf{a}, \mathbf{b} \in \mathbf{C}$, we have that $[\mathbf{a}, \mathbf{b}]_s \subset \mathbf{C}$. Finally, recall that if a closed unit ball $\mathbf{B}^d[\mathbf{c}, 1]$ contains a set $\mathbf{C} \subset \mathbb{E}^d$ and a point $\mathbf{x} \in \mathrm{bd}\mathbf{C}$ is on $S^{d-1}(\mathbf{c}, 1)$, then we say that $S^{d-1}(\mathbf{c}, 1)$ or $\mathbf{B}^d[\mathbf{c}, 1]$ *supports* $\mathbf{C}$ at $\mathbf{x}$. (Here, as usual, the boundary of a set $X \subset \mathbb{E}^d$ is denoted by $\mathrm{bd}X$.)

Theorem 6.6.3 *Let $\mathbf{A} \subset \mathbb{E}^d$ be a closed convex set. Then the following are equivalent.*
(i) $\mathbf{A}$ is spindle convex.
(ii) $\mathbf{A}$ is the intersection of unit balls containing it.
(iii) For every boundary point of $\mathbf{A}$, there is a unit ball that supports $\mathbf{A}$ at that point.

Recall that the interior of a set $X \subset \mathbb{E}^d$ is denoted by $\mathrm{int}X$.

Theorem 6.6.4 *Let* $\mathbf{C}, \mathbf{D} \subset \mathbb{E}^d$ *be spindle convex sets. Suppose* $\mathbf{C}$ *and* $\mathbf{D}$ *have disjoint relative interiors. Then there is a closed unit ball* $\mathbf{B}^d[\mathbf{c}, 1]$ *such that* $\mathbf{C} \subset \mathbf{B}^d[\mathbf{c}, 1]$ *and* $\mathbf{D} \subset \mathbb{E}^d \setminus \mathbf{B}^d(\mathbf{c}, 1)$.

Furthermore, if $\mathbf{C}$ *and* $\mathbf{D}$ *have disjoint closures and one, say* $\mathbf{C}$, *is different from a unit ball, then there is an open unit ball* $\mathbf{B}^d(\mathbf{c}, 1)$ *such that* $\mathbf{C} \subset \mathbf{B}^d(\mathbf{c}, 1)$ *and* $\mathbf{D} \subset \mathbb{E}^d \setminus \mathbf{B}^d[\mathbf{c}, 1]$.

6.7 Carathéodory- and Steinitz-Type Results

In this section we study the spindle convex hull of a set and give analogues of the well-known theorems of Carathéodory and Steinitz to spindle convexity. Carathéodory's theorem (see, e.g., [22]) states that the convex hull of a set $X \subset \mathbb{E}^d$ is the union of simplices with vertices in X. Steinitz's theorem (see, e.g., [22]) is that if a point is in the interior of the convex hull of a set $X \subset \mathbb{E}^d$, then it is also in the interior of the convex hull of at most $2d$ points of X. This number $2d$ cannot be reduced as shown by the cross-polytope and its center point. Recall the following definition introduced in [69]. Let X be a set in $\mathbb{E}^d$. Then the *spindle convex hull* of X is the set defined by $\mathrm{conv}_s X := \bigcap \{C \subset \mathbb{E}^d | X \subset C$ and C is spindle convex in $\mathbb{E}^d\}$. Based on this, we can now phrase the major result of this section proved in [69].

Theorem 6.7.1 *Let* $X \subset \mathbb{E}^d$ *be a closed set.*
(i) If $\mathbf{y} \in \mathrm{bd}(\mathrm{conv}_s X)$, *then there is a set* $\{\mathbf{x}_1, \mathbf{x}_2, \dots, \mathbf{x}_d\} \subset X$ *such that* $\mathbf{y} \in \mathrm{conv}_s\{\mathbf{x}_1, \mathbf{x}_2, \dots, \mathbf{x}_d\}$.
(ii) If $\mathbf{y} \in \mathrm{int}(\mathrm{conv}_s X)$, *then there is a set* $\{\mathbf{x}_1, \mathbf{x}_2, \dots, \mathbf{x}_{d+1}\} \subset X$ *such that* $\mathbf{y} \in \mathrm{int}(\mathrm{conv}_s\{\mathbf{x}_1, \mathbf{x}_2, \dots, \mathbf{x}_{d+1}\})$.

6.8 Illumination of Ball-Polyhedra

Recall the Boltyanski–Hadwiger Illumination Conjecture [155], [78]. Let $\mathbf{K}$ be a convex body (i.e. a compact convex set with nonempty interior) in the d-dimensional Euclidean space $\mathbb{E}^d$, $d \geq 2$. According to Boltyanski [78] the direction $\mathbf{v} \in \mathbb{S}^{d-1}$ (i.e. the unit vector $\mathbf{v}$ of $\mathbb{E}^d$) illuminates the boundary point $\mathbf{b}$ of $\mathbf{K}$ if the halfline emanating from $\mathbf{b}$ having direction vector $\mathbf{v}$ intersects the interior of $\mathbf{K}$, where $\mathbb{S}^{d-1} \subset \mathbb{E}^d$ denotes the $(d-1)$-dimensional unit sphere centered at the origin $\mathbf{o}$ of $\mathbb{E}^d$. Furthermore, the directions $\mathbf{v}_1, \mathbf{v}_2, \dots, \mathbf{v}_n$ illuminate $\mathbf{K}$ if each boundary point of $\mathbf{K}$ is illuminated by at least one of the directions $\mathbf{v}_1, \mathbf{v}_2, \dots, \mathbf{v}_n$. Finally, the smallest n for which there exist n directions that illuminate $\mathbf{K}$ is called the *illumination number* of $\mathbf{K}$ denoted by $I(\mathbf{K})$. An equivalent but somewhat different looking concept of illumination was introduced by Hadwiger in [155]. There he proposed to use point sources instead of directions for the illumination of convex bodies. Based on these circumstances the following conjecture, that was independently raised by

Boltyanski [78] and Hadwiger [155] in 1960, is called the *Boltyanski–Hadwiger Illumination Conjecture*: The illumination number $I(\mathbf{K})$ of any convex body $\mathbf{K}$ in $\mathbb{E}^d$, is at most 2^d and $I(\mathbf{K}) = 2^d$ if and only if $\mathbf{K}$ is an affine d-cube. As discussed in previous sections, this conjecture is proved for $d = 2$ and it is open for all $d \geq 3$ despite the large number of partial results presently known. The following, rather basic principle, can be quite useful for estimating the illumination numbers of some convex bodies in particular, in low dimensions. (It can be proved in a way similar to that of the proof of Theorem 3.5.2.)

Theorem 6.8.1 *Let* $\mathbf{K} \subset \mathbb{E}^d$, $d \geq 3$ *be a convex body and let r be a positive real number with the property that the Gauss image $\nu(F)$ of any face F of $\mathbf{K}$ can be covered by a spherical ball of radius r in $\mathbb{S}^{d-1}$. Moreover, assume that there exist N points of $\mathbb{S}^{d-1}$ with covering radius R satisfying the inequality $r + R \leq \frac{\pi}{2}$. Then $I(\mathbf{K}) \leq N$.*

Using Theorem 6.8.1 as well as the optimal codes for the covering radii of 4 and 5 points on $\mathbb{S}^2$ ([142]) one can prove the first and the second inequality of the theorem stated below. The third inequality has been proved in [69].

Theorem 6.8.2 *Let* $\mathbf{B}[X]$ *be a ball-polyhedron in* $\mathbb{E}^3$, *which is the intersection of the closed 3-dimensional unit balls centered at the points of $X \subset \mathbb{E}^3$.*
(i) If the Euclidean diameter $\mathrm{diam}(X)$ of X satisfies $0 < \mathrm{diam}(X) \leq 0.577$, then $I(\mathbf{B}[X]) = 4$;
(ii) If $\mathrm{diam}(X)$ satisfies $0.577 < \mathrm{diam}(X) \leq 0.774$, then $I(\mathbf{B}[X]) \leq 5$;
(iii) If $0.774 < \mathrm{diam}(X) \leq 1$, then $I(\mathbf{B}[X]) \leq 6$.

By taking a closer look of Schramm's proof [226] of Theorem 3.4.3 and making the necessary modifications, it turns out, that the estimate of Theorem 3.4.3 can be somewhat improved, but more importantly it can be extended to the following family of convex bodies that is much larger than the family of convex bodies of constant width and also includes the family of "fat" ball-polyhedra. Thus, we have the following theorem.

Theorem 6.8.3 *Let* $X \subset \mathbb{E}^d$, $d \geq 3$ *be an arbitrary compact set with $\mathrm{diam}(X) \leq 1$ and let $\mathbf{B}[X]$ be the intersection of the closed d-dimensional unit balls centered at the points of X. Then*

$$I(\mathbf{B}[X]) < 4\left(\frac{\pi}{3}\right)^{\frac{1}{2}} d^{\frac{3}{2}}(3 + \ln d)\left(\frac{3}{2}\right)^{\frac{d}{2}} < 5d^{\frac{3}{2}}(4 + \ln d)\left(\frac{3}{2}\right)^{\frac{d}{2}}.$$

On the one hand, $4\left(\frac{\pi}{3}\right)^{\frac{1}{2}} d^{\frac{3}{2}}(3 + \ln d)\left(\frac{3}{2}\right)^{\frac{d}{2}} < 2^d$ for all $d \geq 15$. (Moreover, for every $\epsilon > 0$ if d is sufficiently large, then $I(\mathbf{B}[X]) < \left(\sqrt{1.5} + \epsilon\right)^d = (1.224\ldots + \epsilon)^d$.) On the other hand, based on the elegant construction of Kahn and Kalai [176], it is known (see [2]), that if d is sufficiently large, then there exists a finite subset X'' of $\{0, 1\}^d$ in $\mathbb{E}^d$ such that any partition of X'' into parts of smaller diameter requires more than $(1.2)^{\sqrt{d}}$ parts. Let X' be the (positive) homothetic copy of X'' having unit diameter and let X be the (not necessarily unique) convex body of constant width one containing X'. Then it follows via standard arguments that $I(\mathbf{B}[X]) > (1.2)^{\sqrt{d}}$ with $X = \mathbf{B}[X]$. The natural question whether there exist "fat" ball-polyhedra with the same property remains open.

Theorem 6.8.2 and Theorem 6.8.3 suggest attacking the Boltyanski–Hadwiger Illumination Conjecture by letting $0 < \mathrm{diam}(X) < 2$ to get arbitrarily close to 2 with circumradius $0 < \mathrm{cr}(X) < 1$.

6.9 The Euler–Poincaré Formula for Ball-Polyhedra

The main result of this section, published in [69], is an Euler–Poincaré-type formula for a large family of ball-polyhedra of $\mathbb{E}^d$ called *standard ball-polyhedra*. This family of ball-polyhedra is an extension of the relevant 3-dimensional family of standard ball-polyhedra already discussed in previous sections. The details are as follows.

Let $S^l(\mathbf{p}, r)$ be a sphere of $\mathbb{E}^d$. The intersection of $S^l(\mathbf{p}, r)$ with an affine subspace of $\mathbb{E}^d$ that passes through $\mathbf{p}$ is called a *great-sphere* of $S^l(\mathbf{p}, r)$. Note that $S^l(\mathbf{p}, r)$ is a great-sphere of itself. Moreover, any great-sphere is itself a sphere. Next, let $\mathbf{P} \subset \mathbb{E}^d$ be a ball-polyhedron with the family of *generating balls* $\mathbf{B}^d[\mathbf{x}_1, 1], \ldots, \mathbf{B}^d[\mathbf{x}_k, 1]$ (meaning that $\mathbf{P} = \cap_{i=1}^k \mathbf{B}^d[\mathbf{x}_i, 1]$). Also, recall that by definition removing any of the balls in question yields that the intersection of the remaining balls becomes a set larger than $\mathbf{P}$. The boundary of a generating ball of $\mathbf{P}$ is called a *generating sphere* of $\mathbf{P}$. A *supporting sphere* $S^l(\mathbf{p}, r)$ of $\mathbf{P}$ is a sphere of dimension l, where $0 \leq l \leq d - 1$, which can be obtained as an intersection of some of the generating spheres of $\mathbf{P}$ such that $\mathbf{P} \cap S^l(\mathbf{p}, r) \neq \emptyset$. Note that the intersection of finitely many spheres in $\mathbb{E}^d$ is either empty, or a sphere, or a point. In the same way that the faces of a convex polytope can be described in terms of supporting affine subspaces, we describe the faces of a certain class of ball-polyhedra in terms of supporting spheres. Thus, let $\mathbf{P}$ be a d-dimensional ball-polyhedron. We say that $\mathbf{P}$ is *standard* if for any supporting sphere $S^l(\mathbf{p}, r)$ of $\mathbf{P}$ the intersection $F := \mathbf{P} \cap S^l(\mathbf{p}, r)$ is homeomorphic to a closed Euclidean ball of some dimension. We call F a *face* of $\mathbf{P}$; the *dimension* of F is the dimension of the ball to which F is homeomorphic. If the dimension is $0, 1$, or $d-1$, then we call the face a *vertex*, an *edge*, or a *facet*, respectively. Note that the dimension of F is independent of the choice of the supporting sphere containing F. The following theorem has been proved in [69], the last part of which is the desired Euler–Poincaré formula for standard ball-polyhedra.

Theorem 6.9.1 *Let Λ be the set containing all faces of a standard ball-polyhedron $\mathbf{P} \subset \mathbb{E}^d$ and the empty set and $\mathbf{P}$ itself. Then Λ is a finite bounded lattice with respect to ordering by inclusion. The atoms of Λ are the vertices of $\mathbf{P}$ and Λ is atomic; that is, for every element $F \in \Lambda$ with $F \neq \emptyset$ there is* *a vertex* $\mathbf{v}$ *of* $\mathbf{P}$ *such that* $\mathbf{v} \in F$. *Moreover,* $\mathbf{P}$ *has* k-*dimensional faces for every* $0 \le k \le d-1$ *and* $\mathbf{P}$ *is the spindle convex hull of its* $(d-2)$-*dimensional faces. Furthermore, no standard ball-polyhedron in* $\mathbb{E}^d$ *is the spindle convex hull of its* $(d-3)$-*dimensional faces. Finally, if* $f_i(\mathbf{P})$ *denotes the number of* i-*dimensional faces of* $\mathbf{P}$, *then*

$$1 + (-1)^{d+1} = \sum_{i=0}^{d-1} (-1)^i f_i(\mathbf{P}).$$

Part II

Selected Proofs

7

Selected Proofs on Sphere Packings

7.1 Proof of Theorem 1.3.5

7.1.1 A proof by estimating the surface area of unions of balls

Let $\mathbf{B}$ denote the unit ball centered at the origin $\mathbf{o}$ of $\mathbb{E}^3$ and let $\mathcal{P} :=$ $\{\mathbf{c}_1 + \mathbf{B}, \mathbf{c}_2 + \mathbf{B}, \dots, \mathbf{c}_n + \mathbf{B}\}$ denote the packing of n unit balls with centers $\mathbf{c}_1, \mathbf{c}_2, \dots, \mathbf{c}_n$ in $\mathbb{E}^3$ having the largest number $C(n)$ of touching pairs among all packings of n unit balls in $\mathbb{E}^3$. ($\mathcal{P}$ might not be uniquely determined up to congruence in which case $\mathcal{P}$ stands for any of those extremal packings.) First, observe that Theorem 1.4.1 and Theorem 2.4.3 imply the following inequality in a straightforward way.

Lemma 7.1.1

$$\frac{n\mathrm{vol}_3(\mathbf{B})}{\mathrm{vol}_3(\bigcup_{i=1}^{n} \mathbf{c}_i + 2\mathbf{B})} \leq \delta(\mathbf{B}) = \frac{\pi}{\sqrt{18}}.$$

Second, the well-known isoperimetric inequality [97] yields the following.

Lemma 7.1.2

$$36\pi\mathrm{vol}_3^2\left(\bigcup_{i=1}^{n} \mathbf{c}_i + 2\mathbf{B}\right) \leq \mathrm{svol}_2^3\left(\mathrm{bd}\left(\bigcup_{i=1}^{n} \mathbf{c}_i + 2\mathbf{B}\right)\right).$$

Thus, Lemma 7.1.1 and Lemma 7.1.2 generate the following inequality.

Corollary 7.1.3

$$4(18\pi)^{\frac{1}{3}} n^{\frac{2}{3}} \leq \mathrm{svol}_2\left(\mathrm{bd}\left(\bigcup_{i=1}^{n} \mathbf{c}_i + 2\mathbf{B}\right)\right).$$

Now, assume that $\mathbf{c}_i + \mathbf{B} \in \mathcal{P}$ is tangent to $\mathbf{c}_j + \mathbf{B} \in \mathcal{P}$ for all $j \in T_i$, where $T_i \subset \{1, 2, \dots, n\}$ stands for the family of indices $1 \leq j \leq n$ for which $\|\mathbf{c}_i - \mathbf{c}_j\| = 2$. Then let $S_i := \mathrm{bd}(\mathbf{c}_i + 2\mathbf{B})$ and let $C_{S_i}(\mathbf{c}_j, \frac{\pi}{6})$ denote the open

K. Bezdek, *Classical Topics in Discrete Geometry*, CMS Books in Mathematics, DOI 10.1007/978-1-4419-0600-7_7, © Springer Science+Business Media, LLC 2010

spherical cap of S_i centered at $\mathbf{c}_j \in S_i$ having angular radius $\frac{\pi}{6}$. Clearly, the family $\{C_{S_i}(\mathbf{c}_j, \frac{\pi}{6}), j \in T_i\}$ consists of pairwise disjoint open spherical caps of S_i; moreover,

$$\frac{\sum_{j \in T_i} \text{svol}_2\left(C_{S_i}(\mathbf{c}_j, \frac{\pi}{6})\right)}{\text{svol}_2\left(\cup_{j \in T_i} C_{S_i}(\mathbf{c}_j, \frac{\pi}{3})\right)} = \frac{\sum_{j \in T_i} \text{Sarea}\left(C(\mathbf{u}_{ij}, \frac{\pi}{6})\right)}{\text{Sarea}\left(\cup_{j \in T_i} C(\mathbf{u}_{ij}, \frac{\pi}{3})\right)}, \tag{7.1}$$

where $\mathbf{u}_{ij} := \frac{1}{2}(\mathbf{c}_j - \mathbf{c}_i) \in \mathbb{S}^2$ and $C(\mathbf{u}_{ij}, \frac{\pi}{6}) \subset \mathbb{S}^2$ (resp., $C(\mathbf{u}_{ij}, \frac{\pi}{3}) \subset \mathbb{S}^2$) denotes the open spherical cap of $\mathbb{S}^2$ centered at $\mathbf{u}_{ij}$ having angular radius $\frac{\pi}{6}$ (resp., $\frac{\pi}{3}$) and where $\text{svol}_2(\cdot)$ (resp., $\text{Sarea}(\cdot)$) denotes the 2-dimensional surface volume measure in $\mathbb{E}^3$ (resp., spherical area measure on $\mathbb{S}^2$) of the corresponding set. Now, Molnár's density bound (see Satz 1 in [200]) implies that

$$\frac{\sum_{j \in T_i} \text{Sarea}\left(C(\mathbf{u}_{ij}, \frac{\pi}{6})\right)}{\text{Sarea}\left(\cup_{j \in T_i} C(\mathbf{u}_{ij}, \frac{\pi}{3})\right)} < 0.89332 . \tag{7.2}$$

In order to estimate $\text{svol}_2\left(\text{bd}\left(\cup_{i=1}^n \mathbf{c}_i + 2\mathbf{B}\right)\right)$ from above let us assume that m members of $\mathcal{P}$ have 12 touching neighbours in $\mathcal{P}$ and k members of $\mathcal{P}$ have at most 9 touching neighbours in $\mathcal{P}$. Thus, $n - m - k$ members of $\mathcal{P}$ have either 10 or 11 touching neighbours in $\mathcal{P}$. Without loss of generality we may assume that $4 \leq k \leq n - m$. Based on the notation just introduced, it is rather easy to see, that (7.1) and (7.2) together with the well-known fact that the kissing number of $\mathbf{B}$ is 12, imply the following estimate.

Corollary 7.1.4

$$\text{svol}_2\left(\text{bd}\left(\bigcup_{i=1}^n \mathbf{c}_i + 2\mathbf{B}\right)\right) < 12.573(n - m - k) + 38.9578k$$

$$< \frac{38.9578}{3}(n - m - k) + 38.9578k .$$

Hence, Corollary 7.1.3 and Corollary 7.1.4 yield in a straightforward way that

$$1.1822n^{\frac{2}{3}} - 3k < n - m - k . \tag{7.3}$$

Finally, as the number $C(n)$ of touching pairs in $\mathcal{P}$ is obviously at most

$$\frac{1}{2}\left(12n - (n - m - k) - 3k\right) ,$$

therefore (7.3) implies that

$$C(n) \leq \frac{1}{2}\left(12n - (n - m - k) - 3k\right) < 6n - 0.5911n^{\frac{2}{3}} < 6n - 0.59n^{\frac{2}{3}},$$

finishing the proof of Theorem 1.3.5.

7.1.2 On the densest packing of congruent spherical caps of special radius

We feel that it is worth making the following comment: it is likely that (7.2) can be replaced by the following sharper estimate.

Conjecture 7.1.5

$$\frac{\sum_{j \in T_i} \text{Sarea}\left(C(\mathbf{u}_{ij}, \frac{\pi}{6})\right)}{\text{Sarea}\left(\cup_{j \in T_i} C(\mathbf{u}_{ij}, \frac{\pi}{3})\right)} \leq 6\left(1 - \frac{\sqrt{3}}{2}\right) = 0.8038\ldots ,$$

with equality when 12 *spherical caps of angular radius* $\frac{\pi}{6}$ *are packed on* $\mathbb{S}^2$.

If so, then one can improve Theorem 1.3.5 as follows.

Proposition 7.1.6 *Conjecture 7.1.5 implies that*

$$C(n) \leq 6n - \frac{3(18\pi)^{\frac{1}{3}}}{2\pi}n^{\frac{2}{3}} = 6n - 1.8326\ldots n^{\frac{2}{3}} .$$

Proof: Indeed, Conjecture 7.1.5 implies in a straightforward way that

$$\text{svol}_2\left(\text{bd}\left(\bigcup_{i=1}^{n} \mathbf{c}_i + 2\mathbf{B}\right)\right)$$

$$\leq 16\pi n - \frac{1}{6\left(1 - \frac{\sqrt{3}}{2}\right)}16\pi\left(1 - \frac{\sqrt{3}}{2}\right)C(n) = 16\pi n - \frac{8\pi}{3}C(n) .$$

The above inequality combined with Corollary 7.1.3 yields

$$4(18\pi)^{\frac{1}{3}}n^{\frac{2}{3}} \leq 16\pi n - \frac{8\pi}{3}C(n) ,$$

from which the inequality of Proposition 7.1.6 follows. □

7.2 Proof of Theorem 1.4.7

7.2.1 The Voronoi star of a Voronoi cell in unit ball packings

Without loss of generality we may assume that the d-dimensional unit ball $\mathbf{B} \subset \mathbb{E}^d$ centered at the origin $\mathbf{o}$ of $\mathbb{E}^d$ is one of the unit balls of the given unit ball packing in $\mathbb{E}^d, d \geq 2$. Let $\mathbf{V}$ be the Voronoi cell assigned to $\mathbf{B}$. We may assume that $\mathbf{V}$ is bounded; that is, $\mathbf{V}$ is a d-dimensional convex polytope in $\mathbb{E}^d$.

First, following [218], we dissect $\mathbf{V}$ into finitely many d-dimensional simplices as follows. Let F_i denote an arbitrary i-dimensional face of $\mathbf{V}$,

$0 \leq i \leq d - 1$. Let the chain $F_0 \subset F_1 \subset \cdots \subset F_{d-1}$ be called a *flag* of $\mathbf{V}$, and let $\mathcal{F}$ be the family of all flags of $\mathbf{V}$. Now, let $f \in \mathcal{F}$ be an arbitrary flag of $\mathbf{V}$ with the associated chain $F_0 \subset F_1 \subset \cdots \subset F_{d-1}$. Then let $\mathbf{v}_i \in F_{d-i}$ be the point of F_{d-i} closest to $\mathbf{o}$, $1 \leq i \leq d$. Finally, let $\mathbf{V}_f := \mathrm{conv}\{\mathbf{o}, \mathbf{v}_1, \ldots, \mathbf{v}_d\}$, where $\mathrm{conv}(\cdot)$ stands for the convex hull of the given set. It is easy to see that the family $\mathcal{V} := \{\mathbf{V}_f \mid f \in \mathcal{F} \text{ and } \dim(\mathbf{V}_f) = d\}$ of d-dimensional simplices forms a tiling of $\mathbf{V}$ (i.e., $\cup_{\mathbf{V}_f \in \mathcal{V}} \mathbf{V}_f = \mathbf{V}$ and no two simplices of $\mathcal{V}$ have an interior point in common). This tiling is a rather special one, namely the d-dimensional simplices of $\mathcal{V}$ have $\mathbf{o}$ as a common vertex; moreover the union of their facets opposite to $\mathbf{o}$ is the boundary $\mathrm{bd}\mathbf{V}$ of $\mathbf{V}$. Finally, as shown in [218], for any $\mathbf{V}_f \in \mathcal{V}$ with $\mathbf{V}_f = \mathrm{conv}\{\mathbf{o}, \mathbf{v}_1, \ldots, \mathbf{v}_d\}$ we have that

$$\sqrt{\frac{2i}{i+1}} \leq \|\mathbf{v}_i\| = \mathrm{dist}\left(\mathbf{o}, \mathrm{conv}\{\mathbf{v}_i, \mathbf{v}_{i+1}, \ldots, \mathbf{v}_d\}\right), 1 \leq i \leq d, \qquad (7.4)$$

where $\mathrm{dist}(\cdot, \cdot)$ (resp., $\|\cdot\|$) stands for the Euclidean distance function (resp., norm) in $\mathbb{E}^d$.

Second, we define the *Voronoi star* $\mathbf{V}^* \subset \mathbf{V}$ assigned to the Voronoi cell $\mathbf{V}$ as follows. Let $\mathbf{V}_f \in \mathcal{V}$ with $\mathbf{V}_f = \mathrm{conv}\{\mathbf{o}, \mathbf{v}_1, \ldots, \mathbf{v}_d\}$. Then let $\mathbf{v}_1^* := H \cap \mathrm{lin}\{\mathbf{v}_1\}$, where H denotes the hyperplane parallel to the hyperplane $\mathrm{aff}\{\mathbf{v}_1, \ldots, \mathbf{v}_d\}$ and tangent to $\mathbf{B}$ such that it separates $\mathbf{o}$ from $\mathrm{aff}\{\mathbf{v}_1, \ldots, \mathbf{v}_d\}$ (with $\mathrm{lin}(\cdot)$ and $\mathrm{aff}(\cdot)$ standing for the linear and affine hulls of the given sets in $\mathbb{E}^d$). Finally, let $\mathbf{V}_f^* := \mathrm{conv}\{\mathbf{o}, \mathbf{v}_1^*, \mathbf{v}_2, \ldots, \mathbf{v}_d\}$ and let the Voronoi star $\mathbf{V}^*$ of $\mathbf{V}$ be defined as $\mathbf{V}^* := \cup_{\mathbf{V}_f \in \mathcal{V}} \mathbf{V}_f^*$. It follows from the definition of the Voronoi star and from (7.4) that the following inequalities and (surface) volume formula hold:

$$1 \leq \|\mathbf{v}_1^*\| = \mathrm{dist}\left(\mathbf{o}, \mathrm{conv}\{\mathbf{v}_1^*, \mathbf{v}_2, \ldots, \mathbf{v}_d\}\right) \leq \|\mathbf{v}_1\|, \qquad (7.5)$$

$$\sqrt{\frac{2i}{i+1}} \leq \|\mathbf{v}_i\| = \mathrm{dist}\left(\mathbf{o}, \mathrm{conv}\{\mathbf{v}_i, \mathbf{v}_{i+1}, \ldots, \mathbf{v}_d\}\right), 2 \leq i \leq d, \text{ and} \qquad (7.6)$$

$$\mathrm{vol}_d(\mathbf{V}^*) = \frac{1}{d}\mathrm{svol}_{d-1}(\mathrm{bd}\mathbf{V}), \qquad (7.7)$$

where $\mathrm{vol}_d(\cdot)$ (resp., $\mathrm{svol}_{d-1}(\cdot)$) refers to the d-dimensional (resp., $(d-1)$-dimensional) volume (resp., surface volume) measure.

7.2.2 Estimating the volume of a Voronoi star from below

As an obvious corollary of (7.7), we find that Theorem 1.4.7 follows from the following theorem.

Theorem 7.2.1 $\mathrm{vol}_d(\mathbf{V}^*) \geq \frac{\omega_d}{\sigma_d}$.

Proof: The main tool of our proof is the following lemma of Rogers. (See [218] and [219] for the original version of the lemma, which is somewhat different from the equivalent version below. Also, for a strengthening we refer the interested reader to Lemma 7.3.11.)

Lemma 7.2.2 *Let* $\mathbf{W} := \operatorname{conv}\{\mathbf{o}, \mathbf{w}_1, \ldots, \mathbf{w}_d\}$ *be a d-dimensional simplex of* $\mathbb{E}^d$ *having the property that* $\operatorname{lin}\{\mathbf{w}_j - \mathbf{w}_i \mid i < j \leq d\}$ *is orthogonal to the vector* $\mathbf{w}_i$ *in* $\mathbb{E}^d$ *for all* $1 \leq i \leq d-1$ *(i.e., let* $\mathbf{W}$ *be a d-dimensional orthoscheme in* $\mathbb{E}^d$*). Moreover, let* $\mathbf{U} := \operatorname{conv}\{\mathbf{o}, \mathbf{u}_1, \ldots, \mathbf{u}_d\}$ *be a d-dimensional simplex of* $\mathbb{E}^d$ *such that* $\|\mathbf{u}_i\| = \operatorname{dist}(\mathbf{o}, \operatorname{conv}\{\mathbf{u}_i, \mathbf{u}_{i+1}, \ldots, \mathbf{u}_d\})$ *for all* $1 \leq i \leq d$*. If* $\|\mathbf{w}_i\| \leq \|\mathbf{u}_i\|$ *holds for all* $1 \leq i \leq d$*, then*

$$\frac{\operatorname{vol}_d(\mathbf{W})}{\operatorname{vol}_d(\mathbf{B} \cap \mathbf{W})} \leq \frac{\operatorname{vol}_d(\mathbf{U})}{\operatorname{vol}_d(\mathbf{B} \cap \mathbf{U})},$$

where $\mathbf{B}$ *stands for the d-dimensional unit ball centered at the origin* $\mathbf{o}$ *of* $\mathbb{E}^d$*.*

Now, let $\mathbf{W}$ be the orthoscheme of Lemma 7.2.2 with the additional property that $\|\mathbf{w}_i\| = \sqrt{\frac{2i}{i+1}}$ for all $1 \leq i \leq d$. Notice that a regular d-dimensional simplex of edge length 2 in $\mathbb{E}^d$ can be dissected into $(d+1)!$ d-dimensional simplices, each congruent to $\mathbf{W}$. This implies that

$$\sigma_d = \frac{\operatorname{vol}_d(\mathbf{B} \cap \mathbf{W})}{\operatorname{vol}_d(\mathbf{W})}. \tag{7.8}$$

Finally, let $\mathbf{U} := \mathbf{V}_f^* = \operatorname{conv}\{\mathbf{o}, \mathbf{v}_1^*, \mathbf{v}_2, \ldots, \mathbf{v}_d\}$ for $\mathbf{V}_f \in \mathcal{V}$. Clearly, (7.5) and (7.6) show that $\mathbf{W}$ and $\mathbf{U}$, just introduced, satisfy the assumptions of Lemma 7.2.2. Thus, Lemma 7.2.2 and (7.8) imply that

$$\frac{1}{\sigma_d} \leq \frac{\operatorname{vol}_d(\mathbf{V}_f^*)}{\operatorname{vol}_d(\mathbf{B} \cap \mathbf{V}_f^*)}. \tag{7.9}$$

Hence, (7.9) yields that

$$\frac{\omega_d}{\sigma_d} \leq \sum_{\mathbf{V}_f \in \mathcal{V}} \operatorname{vol}_d(\mathbf{B} \cap \mathbf{V}_f^*) \frac{\operatorname{vol}_d(\mathbf{V}_f^*)}{\operatorname{vol}_d(\mathbf{B} \cap \mathbf{V}_f^*)} = \sum_{\mathbf{V}_f \in \mathcal{V}} \operatorname{vol}_d(\mathbf{V}_f^*) = \operatorname{vol}_d(\mathbf{V}^*),$$

finishing the proof of Theorem 7.2.1. □

7.3 Proof of Theorem 1.4.8

7.3.1 Basic metric properties of Voronoi cells in unit ball packings

Let $\mathbf{P}$ be a bounded Voronoi cell, that is, a d-dimensional Voronoi polytope of a packing $\mathcal{P}$ of d-dimensional unit balls in $\mathbb{E}^d$. Without loss of generality we may assume that the unit ball $\mathbf{B} = \{\mathbf{x} \in \mathbb{E}^d | \operatorname{dist}(\mathbf{o}, \mathbf{x}) = \|\mathbf{x}\| \leq 1\}$ centered at the origin $\mathbf{o}$ of $\mathbb{E}^d$ is one of the unit balls of $\mathcal{P}$ with $\mathbf{P}$ as its Voronoi cell. Then $\mathbf{P}$ is the intersection of finitely many closed halfspaces of $\mathbb{E}^d$ each of which is bounded by a hyperplane that is the perpendicular bisector of a line segment $\mathbf{ox}$ with $\mathbf{x}$ being the center of some unit ball of $\mathcal{P}$. Now, let F_{d-i} be an arbitrary $(d-i)$-dimensional face of $\mathbf{P}$, $1 \leq i \leq d$. Then clearly there are at least $i+1$ Voronoi cells of $\mathcal{P}$ which meet along the face F_{d-i}, that is, contain F_{d-i} (one of which is, of course, $\mathbf{P}$). Also, it is clear from the construction that the affine hull of centers of the unit balls sitting in all of these Voronoi cells is orthogonal to $\operatorname{aff} F_{d-i}$. Thus, there are unit balls of these Voronoi cells with centers $\{\mathbf{o}, \mathbf{x}_1, \ldots, \mathbf{x}_i\}$ such that $X = \operatorname{conv}\{\mathbf{o}, \mathbf{x}_1, \ldots, \mathbf{x}_i\}$ is an i-dimensional simplex and of course, $\operatorname{aff} X$ is orthogonal to $\operatorname{aff} F_{d-i}$. Hence, if $R(F_{d-i})$ denotes the radius of the $(i-1)$-dimensional sphere that passes through the vertices of X, then

$$R(F_{d-i}) = \operatorname{dist}(\mathbf{o}, \operatorname{aff} F_{d-i}), \text{ where } 1 \leq i \leq d.$$

As the following statements are well known and their proofs are relatively straightforward, we refer the interested reader to the relevant section in [56] for the details of those proofs.

Lemma 7.3.1 *If $F_{d-i-1} \subset F_{d-i}$ and $R(F_{d-i}) = R < \sqrt{2}$ for some $i, 1 \leq i \leq d - 1$, then*

$$\frac{2}{\sqrt{4 - R^2}} \leq R(F_{d-i-1}).$$

Corollary 7.3.2 $\sqrt{\frac{2i}{i+1}} \leq R(F_{d-i})$ *for all $1 \leq i \leq d$.*

Lemma 7.3.3 *If $R(F_{d-i}) < \sqrt{2}$ for some $i, 1 \leq i \leq d$, then the orthogonal projection of $\mathbf{o}$ onto $\operatorname{aff} F_{d-i}$ belongs to $\operatorname{relint} F_{d-i}$ and so $R(F_{d-i}) = \operatorname{dist}(\mathbf{o}, F_{d-i})$.*

7.3.2 Wedges of types I, II, and III, and truncated wedges of types I, and II

Let $F_0 \subset F_1 \subset \cdots \subset F_{d-1}$ be an arbitrary flag of the Voronoi polytope $\mathbf{P}$. Then let $\mathbf{r}_i \in F_{d-i}$ be the uniquely determined point of the $(d-i)$-dimensional face F_{d-i} of $\mathbf{P}$ that is closest to the center point $\mathbf{o}$ of $\mathbf{P}$; that is, let

$$\mathbf{r}_i \in F_{d-i} \text{ such that } \|\mathbf{r}_i\| = \min\{\|\mathbf{x}\| \mid \mathbf{x} \in F_{d-i}\}, \text{ where } 1 \leq i \leq d.$$

Definition 7.3.4 *If the vectors $\mathbf{r}_1, \ldots, \mathbf{r}_i$ are linearly independent in $\mathbb{E}^d$, then we call $\operatorname{conv}\{\mathbf{o}, \mathbf{r}_1, \ldots, \mathbf{r}_i\}$ the i-dimensional Rogers simplex assigned to the subflag $F_{d-i} \subset \cdots \subset F_{d-1}$ of the Voronoi polytope $\mathbf{P}$, where $1 \leq i \leq d$. If $\operatorname{conv}\{\mathbf{o}, \mathbf{r}_1, \ldots, \mathbf{r}_d\} \subset \mathbb{E}^d$ is the d-dimensional Rogers simplex assigned to the flag $F_0 \subset \cdots \subset F_{d-1}$ of $\mathbf{P}$, then $\operatorname{conv}\{\mathbf{r}_{d-i}, \ldots, \mathbf{r}_d\}$ is* *called the i-dimensional base of the given d-dimensional Rogers simplex and* $\operatorname{dist}(\mathbf{o}, \operatorname{aff}\{\mathbf{r}_{d-i}, \ldots, \mathbf{r}_d\}) = \operatorname{dist}(\mathbf{o}, \operatorname{aff} F_i) = R(F_i)$ *is called the height assigned to the i-dimensional base, where* $1 \leq i \leq d$.

Definition 7.3.5 *The i-dimensional simplex* $Y = \operatorname{conv}\{\mathbf{o}, \mathbf{y}_1, \ldots, \mathbf{y}_i\} \subset \mathbb{E}^d$ *with vertices* $\mathbf{y}_0 = \mathbf{o}, \mathbf{y}_1, \ldots, \mathbf{y}_i$ *is called an i-dimensional orthoscheme if for each* $j, 0 \leq j \leq i - 1$ *the vector* $\mathbf{y}_j$ *is orthogonal to the linear hull* $\operatorname{lin}\{\mathbf{y}_k - \mathbf{y}_j \mid j + 1 \leq k \leq i\}$, *where* $1 \leq i \leq d$.

It is shown in [218] that the union of the d-dimensional Rogers simplices of the Voronoi polytope $\mathbf{P}$ is the polytope $\mathbf{P}$ itself and their interiors are pairwise disjoint. This fact together with Corollary 7.3.2 and Lemma 7.3.3 imply the following metric properties of Rogers simplices in a straightforward way.

Lemma 7.3.6
(1) If $\operatorname{conv}\{\mathbf{o}, \mathbf{r}_1, \ldots, \mathbf{r}_i\}$ *is an i-dimensional Rogers simplex assigned to the subflag* $F_{d-i} \subset \cdots \subset F_{d-1}$ *of the Voronoi polytope* $\mathbf{P}$, *then* $\sqrt{\frac{2j}{j+1}} \leq \|\mathbf{r}_j\|$ *for all* $1 \leq j \leq i$, *where* $1 \leq i \leq d$.
(2) If $F_{d-i} \subset \cdots \subset F_{d-1}$ *is a subflag of the Voronoi polytope* $\mathbf{P}$ *with* $R(F_{d-i}) < \sqrt{2}$, *then* $\operatorname{conv}\{\mathbf{o}, \mathbf{r}_1, \ldots, \mathbf{r}_i\}$ *is an i-dimensional Rogers simplex which is, in fact, an i-dimensional orthoscheme (in short, an i-dimensional Rogers orthoscheme) with the property that each* $\mathbf{r}_j \in \operatorname{relint} F_{d-j}, 1 \leq j \leq i$ *is the orthogonal projection of* $\mathbf{o}$ *onto* $\operatorname{aff} F_{d-j}$, *where* $1 \leq i \leq d$.
(3) If $F_2 \subset \cdots \subset F_{d-1}$ *is a subflag of the Voronoi polytope* $\mathbf{P} \subset \mathbb{E}^d, 3 \leq d$ *with* $R(F_2) < \sqrt{2}$, *then the union of the 2-dimensional bases of the d-dimensional Rogers simplices that contain the orthoscheme* $\operatorname{conv}\{\mathbf{o}, \mathbf{r}_1, \ldots, \mathbf{r}_{d-2}\}$ *is the (uniquely determined) 2-dimensional face* F_2 *of the Voronoi polytope* $\mathbf{P}$ *that is totally orthogonal to* $\operatorname{conv}\{\mathbf{o}, \mathbf{r}_1, \ldots, \mathbf{r}_{d-2}\}$ *at the point* $\mathbf{r}_{d-2}$ *and so,* $\|\mathbf{r}_{d-2}\| = \operatorname{dist}(\mathbf{o}, \operatorname{aff} F_2)$ *with* $\mathbf{r}_{d-2} \in \operatorname{relint} F_2$.

Now we are ready for the definitions of wedges and truncated wedges. Recall that for any 2-dimensional face F_2 of the Voronoi polytope $\mathbf{P} \subset \mathbb{E}^d, d \geq 3$ we have that $\sqrt{\frac{2(d-2)}{d-1}} \leq R(F_2)$.

Definition 7.3.7
(1) Let F_2 *be a 2−dimensional face of the Voronoi polytope* $\mathbf{P} \subset \mathbb{E}^d, d \geq 3$ *with* $\sqrt{\frac{2(d-2)}{d-1}} \leq R(F_2) < \sqrt{\frac{2(d-1)}{d}}$ *and let* $\operatorname{conv}\{\mathbf{o}, \mathbf{r}_1, \ldots, \mathbf{r}_{d-2}\}$ *be any* $(d-2)$-*dimensional Rogers simplex with* $\mathbf{r}_{d-2} \in \operatorname{relint} F_2$. *Then the union* $\mathbf{W}_I$ *of the d-dimensional Rogers simplices of* $\mathbf{P}$ *that contain the orthoscheme* $\operatorname{conv}\{\mathbf{o}, \mathbf{r}_1, \ldots, \mathbf{r}_{d-2}\}$ *is called a wedge of type I (generated by the* $(d-2)$-*dimensional Rogers orthoscheme* $\operatorname{conv}\{\mathbf{o}, \mathbf{r}_1, \ldots, \mathbf{r}_{d-2}\}$). F_2 *is called the 2-dimensional base of* $\mathbf{W}_I$, *and* $\|\mathbf{r}_{d-2}\| = \operatorname{dist}(\mathbf{o}, \operatorname{aff} F_2)$ *is the height of* $\mathbf{W}_I$ *assigned to the base* F_2.

(2) Let F_2 be a 2-dimensional face of the Voronoi polytope $\mathbf{P} \subset \mathbb{E}^d, d \geq 3$ with $\sqrt{\frac{2(d-1)}{d}} \leq R(F_2) < \sqrt{\frac{2d}{d+1}}$ and let $\mathrm{conv}\{\mathbf{o},\mathbf{r}_1,\ldots,\mathbf{r}_{d-2}\}$ be any $(d-2)$-dimensional Rogers simplex with $\mathbf{r}_{d-2} \in \mathrm{relint}F_2$. Then the union $\mathbf{W}_{II}$ of the d-dimensional Rogers simplices of $\mathbf{P}$ that contain the orthoscheme $\mathrm{conv}\{\mathbf{o},\mathbf{r}_1,\ldots,\mathbf{r}_{d-2}\}$ is called a wedge of type II (generated by the $(d-2)$-dimensional Rogers orthoscheme $\mathrm{conv}\{\mathbf{o},\mathbf{r}_1,\ldots,\mathbf{r}_{d-2}\}$). F_2 is called the 2-dimensional base of $\mathbf{W}_{II}$, and $\|\mathbf{r}_{d-2}\| = \mathrm{dist}(\mathbf{o}, \mathrm{aff}F_2)$ is the height of $\mathbf{W}_{II}$ assigned to the base F_2.

(3) Let $\mathrm{conv}\{\mathbf{o},\mathbf{r}_1,\ldots,\mathbf{r}_d\}$ be the d-dimensional Rogers simplex assigned to the flag $F_0 \subset F_1 \cdots \subset F_{d-1}$ of the Voronoi polytope $\mathbf{P} \subset \mathbb{E}^d, d \geq 3$ with $\sqrt{\frac{2d}{d+1}} \leq R(F_2)$. Then $\mathbf{W}_{III} = \mathrm{conv}\{\mathbf{o},\mathbf{r}_1,\ldots,\mathbf{r}_d\}$ is called a wedge of type III.

At this point, it useful to recall, that for any vertex F_0 of the Voronoi polytope $\mathbf{P} \subset \mathbb{E}^d$ we have that $\sqrt{\frac{2d}{d+1}} \leq R(F_0)$.

Definition 7.3.8 *Let $\overline{\mathbf{B}} = \left\{\mathbf{x} \in \mathbb{E}^d | \mathrm{dist}(\mathbf{o},\mathbf{x}) = \|\mathbf{x}\| \leq \sqrt{\frac{2d}{d+1}}\right\}$.*

(1) If $\mathbf{W}_I$ is a wedge of type I with the 2-dimensional base F_2 which is generated by the $(d-2)$-dimensional Rogers orthoscheme $\mathrm{conv}\{\mathbf{o},\mathbf{r}_1,\ldots,\mathbf{r}_{d-2}\}$ of the Voronoi polytope $\mathbf{P} \subset \mathbb{E}^d, d \geq 3$, then

$$\overline{\mathbf{W}}_I = \mathrm{conv}\left((\overline{\mathbf{B}} \cap F_2) \cup \{\mathbf{o} = \mathbf{r}_0,\ldots,\mathbf{r}_{d-3}\}\right)$$

is called the truncated wedge of type I with the 2-dimensional base $\overline{\mathbf{B}} \cap F_2$ generated by the $(d-2)$-dimensional Rogers orthoscheme

$$\mathrm{conv}\{\mathbf{o},\mathbf{r}_1,\ldots,\mathbf{r}_{d-2}\}.$$

(2) If $\mathbf{W}_{II}$ is a wedge of type II with the 2-dimensional base F_2 which is generated by the $(d-2)$-dimensional Rogers orthoscheme $\mathrm{conv}\{\mathbf{o},\mathbf{r}_1,\ldots,\mathbf{r}_{d-2}\}$ of the Voronoi polytope $\mathbf{P} \subset \mathbb{E}^d, d \geq 3$, then

$$\overline{\mathbf{W}}_{II} = \mathrm{conv}\left((\overline{\mathbf{B}} \cap F_2) \cup \{\mathbf{o} = \mathbf{r}_0,\ldots,\mathbf{r}_{d-3}\}\right)$$

is called the truncated wedge of type II with the 2-dimensional base $\overline{\mathbf{B}} \cap F_2$ generated by the $(d-2)$-dimensional Rogers orthoscheme

$$\mathrm{conv}\{\mathbf{o},\mathbf{r}_1,\ldots,\mathbf{r}_{d-2}\}.$$

As the following claim can be proved by Lemma 7.3.6 in a straightforward way, we leave the relevant details to the reader.

Lemma 7.3.9

(1) Let $\mathbf{W}_I$ *(resp.,* $\mathbf{W}_{II}$*) denote the wedge of type I (resp., of type II) with the 2-dimensional base* F_2 *which is generated by the* $(d-2)$*-dimensional Rogers orthoscheme* $\mathrm{conv}\{\mathbf{o}, \mathbf{r}_1, \ldots, \mathbf{r}_{d-2}\}$ *of the Voronoi polytope* $\mathbf{P} \subset \mathbb{E}^d, d \geq 3$. *If the points* $\mathbf{x}, \mathbf{y} \in \mathrm{aff}\, F_2$ *are chosen so that the triangle* $\triangle \mathbf{r}_{d-2}\mathbf{xy}$ *has a right angle at the vertex* $\mathbf{x}$*, then* $\mathrm{conv}\{\mathbf{o}, \mathbf{r}_1, \ldots, \mathbf{r}_{d-2}, \mathbf{x}, \mathbf{y}\}$ *is a* d*-dimensional orthoscheme. Moreover, if* $\mathbf{z} \in \mathrm{aff}\, F_2$ *is an arbitrary point, then* $\mathrm{conv}\{\mathbf{o} = \mathbf{r}_0, \ldots, \mathbf{r}_{d-3}, \mathbf{z}\}$ *is a* $(d-2)$*-dimensional orthoscheme.*

(2) Let $\mathbf{W}_I$ *denote the wedge of type I with the 2-dimensional base* F_2 *which is generated by the* $(d-2)$*-dimensional Rogers orthoscheme* $\mathrm{conv}\{\mathbf{o} = \mathbf{r}_0, \mathbf{r}_1, \ldots, \mathbf{r}_{d-2}\}$ *of the Voronoi polytope* $\mathbf{P} \subset \mathbb{E}^d, d \geq 3$. *Let* $Q_2 \subset \mathrm{aff}\, F_2$ *and* $Q_2^* \subset \mathrm{aff}\, F_2$ *be compact convex sets with* $\mathrm{relint}\, Q_2 \cap \mathrm{relint}\, Q_2^* = \emptyset$. *If* $K_2 = Q_2$ *(resp.,* $K_2^* = Q_2^*$*) and* $K_j = \mathrm{conv}(K_{j-1} \cup \{\mathbf{r}_{d-j}\})$ *(resp.,* $K_j^* = \mathrm{conv}(K_{j-1}^* \cup \{\mathbf{r}_{d-j}\})$*) for* $j = 3, \ldots, d$*, then* $K_d = \mathrm{conv}(Q_2 \cup \{\mathbf{o} = \mathbf{r}_0, \ldots, \mathbf{r}_{d-3}\})$ *(resp.,* $K_d^* = \mathrm{conv}(Q_2^* \cup \{\mathbf{o} = \mathbf{r}_0, \ldots, \mathbf{r}_{d-3}\})$*), moreover* $\mathrm{relint}\, K_d \cap \mathrm{relint}\, K_d^* = \emptyset$. *A similar statement holds for* $\mathbf{W}_{II}$.

(3) Let $\mathbf{W}_I$ *(resp.,* $\overline{\mathbf{W}}_I$*) denote the wedge of type I (resp., truncated wedge of type I) with the 2-dimensional base* F_2 *(resp.,* $\overline{\mathbf{B}} \cap F_2$*) which is generated by the* $(d-2)$*-dimensional Rogers orthoscheme* $\mathrm{conv}\{\mathbf{o} = \mathbf{r}_0, \mathbf{r}_1, \ldots, \mathbf{r}_{d-2}\}$ *of the Voronoi polytope* $\mathbf{P} \subset \mathbb{E}^d, d \geq 3$. *If* $K_2 = F_2$ *(resp.,* $K_2 = \overline{\mathbf{B}} \cap F_2$*) and* $K_j = \mathrm{conv}(K_{j-1} \cup \{\mathbf{r}_{d-j}\})$ *for* $j = 3, \ldots, d$*, then* $K_d = \mathbf{W}_I$ *(resp.,* $K_d = \overline{\mathbf{W}}_I$*). Similar statements hold for* $\mathbf{W}_{II}$ *and* $\overline{\mathbf{W}}_{II}$.

We close this section with the following important observation published in [56], and refer the interested reader to [56] for the details of the seven-page proof, which is based on Corollary 7.3.2 and Lemma 7.3.3.

Lemma 7.3.10 *Let* $\overline{\mathbf{B}} \cap F_2$ *be the 2-dimensional base of the type I truncated wedge* $\overline{\mathbf{W}}_I$ *(resp., type II truncated wedge* $\overline{\mathbf{W}}_{II}$*) in the Voronoi polytope* $\mathbf{P} \subset \mathbb{E}^d$ *of dimension* $d \geq 8$. *Then the number of line segments of positive length in* $\mathrm{relbd}(\overline{\mathbf{B}} \cap F_2)$ *is at most 4.*

7.3.3 The lemma of comparison and a characterization of regular polytopes

Recall that $\mathbf{B} = \{\mathbf{x} \in \mathbb{E}^d | \mathrm{dist}(\mathbf{o}, \mathbf{x}) = \|\mathbf{x}\| \leq 1\}$ and let

$$S = \{\mathbf{x} \in \mathbb{E}^d | \mathrm{dist}(\mathbf{o}, \mathbf{x}) = \|\mathbf{x}\| = 1\}.$$

Then let $H \subset \mathbb{E}^d$ be a hyperplane disjoint from the interior of the unit ball $\mathbf{B}$ and let $Q \subset H$ be an arbitrary $(d-1)$-dimensional compact convex set. If $[\mathbf{o}, Q]$ denotes the convex cone $\mathrm{conv}(\{\mathbf{o}\} \cup Q)$ with apex $\mathbf{o}$ and base Q, then the *(volume) density* $\delta([\mathbf{o}, Q], \mathbf{B})$ of the unit ball $\mathbf{B}$ in the cone $[\mathbf{o}, Q]$ is defined as

$$\delta([\mathbf{o}, Q], B) = \frac{\mathrm{vol}_d([\mathbf{o}, Q] \cap \mathbf{B})}{\mathrm{vol}_d([\mathbf{o}, Q])},$$

where $\mathrm{vol}_d(\cdot)$ refers to the corresponding d-dimensional Euclidean volume measure. It is natural to introduce the following very similar notion. The *surface density* $\widehat{\delta}([\mathbf{o}, Q], S)$ of the unit sphere S in the convex cone $[\mathbf{o}, Q]$ with apex $\mathbf{o}$ and base Q is defined by

$$\widehat{\delta}([\mathbf{o}, Q], S) = \frac{\mathrm{Svol}_{d-1}([\mathbf{o}, Q] \cap S)}{\mathrm{vol}_{d-1}(Q)},$$

where $\mathrm{Svol}_{d-1}(\cdot)$ refers to the corresponding $(d-1)$-dimensional spherical volume measure.

If $h = \mathrm{dist}(\mathbf{o}, H)$, then clearly $h \cdot \delta([\mathbf{o}, Q], \mathbf{B}) = \widehat{\delta}([\mathbf{o}, Q], S)$. We need the following statement, the first part of which is due to Rogers [218] and the second part of which has been proved by the author in [55].

Lemma 7.3.11 *Let* $\mathbf{U} = \mathrm{conv}\{\mathbf{o}, \mathbf{u}_1, \ldots, \mathbf{u}_d\}$ *be a d-dimensional orthoscheme in* $\mathbb{E}^d$ *and let* $\mathbf{V} = \mathrm{conv}\{\mathbf{o}, \mathbf{v}_1, \ldots, \mathbf{v}_d\}$ *be a d-dimensional simplex of* $\mathbb{E}^d$ *such that* $\|\mathbf{v}_i\| = \mathrm{dist}(\mathbf{o}, \mathrm{conv}\{\mathbf{v}_i, \mathbf{v}_{i+1}, \ldots, \mathbf{v}_d\})$ *for all* $1 \le i \le d-1$. *If* $1 \le \|\mathbf{u}_i\| \le \|\mathbf{v}_i\|$ *holds for all* $1 \le i \le d$, *then*
(1) $\delta(\mathbf{U}, \mathbf{B}) \ge \delta(\mathbf{V}, \mathbf{B})$ *and*
(2) $\widehat{\delta}(\mathbf{U}, S) \ge \widehat{\delta}(\mathbf{V}, S)$.

For the sake of completeness we mention the following statement that follows from Lemma 7.3.11 using the special decomposition of convex polytopes into Rogers simplices. Actually, the characterization of regular polytopes through the corresponding volume (resp., surface volume) inequality below was first observed by Böröczky and Máthéné Bognár [91] (resp., by the author [55]). (In fact, it is easy to see that the statement on surface volume implies the one on volume.) For more details on related problems we refer the interested reader to [93].

Corollary 7.3.12 *Let* $\mathbf{U}'$ *be a regular convex polytope in* $\mathbb{E}^d$ *with circumcenter* $\mathbf{o}$ *and let* s_i *denote the distance of an i-dimensional face of* $\mathbf{U}'$ *from* $\mathbf{o}$, $0 \le i \le d-1$. *If* $\mathbf{V}'$ *is an arbitrary convex polytope in* $\mathbb{E}^d$ *such that* $\mathbf{o} \in \mathrm{int}\mathbf{V}'$ *and the distance of any i-dimensional face of* $\mathbf{V}'$ *from* $\mathbf{o}$ *is at least* s_i *for all* $0 \le i \le d-1$, *then* $\mathrm{vol}_d(\mathbf{V}') \ge \mathrm{vol}_d(\mathbf{U}')$ *(resp.,* $\mathrm{svol}_{d-1}(\mathbf{V}') \ge \mathrm{svol}_{d-1}(\mathbf{U}')$). *Moreover, equality holds if and only if* $\mathbf{V}'$ *is congruent to* $\mathbf{U}'$ *and its circumcenter is* $\mathbf{o}$.

7.3.4 Volume formulas for (truncated) wedges

Definition 7.3.13 *Let* $\mathbf{x}_1, \ldots, \mathbf{x}_n, n \ge 1$ *be points in* $\mathbb{E}^d, d \ge 1$ *and let* $X \subset \mathbb{E}^d$ *be an arbitrary convex set. If* $X_0 = X$ *and* $X_m = \mathrm{conv}(\{\mathbf{x}_{n-(m-1)}\} \cup X_{m-1})$ *for* $m = 1, \ldots, n$, *then we denote the final convex set* X_n *by*

$$[\mathbf{x}_1, \ldots, \mathbf{x}_n, X].$$

Definition 7.3.14 *Let* $\mathbf{W}_I$ *(resp.,* $\overline{\mathbf{W}}_I$ *) denote the wedge (resp., truncated wedge) of type I with the 2-dimensional base* F_2 *(resp.,* $\overline{\mathbf{B}} \cap F_2$ *) which is generated by the* $(d-2)$*-dimensional Rogers orthoscheme* $\mathrm{conv}\{\mathbf{o}, \mathbf{r}_1, \ldots, \mathbf{r}_{d-2}\}$ *of the Voronoi polytope* $\mathbf{P} \subset \mathbb{E}^d, d \geq 4$*. Then let*

$$Q_I = [\mathbf{r}_1, \ldots, \mathbf{r}_{d-3}, F_2] \; (resp., \overline{Q}_I = [\mathbf{r}_1, \ldots, \mathbf{r}_{d-3}, \overline{\mathbf{B}} \cap F_2])$$

be called the $(d-1)$*-dimensional base of the type I wedge* $\mathbf{W}_I = [\mathbf{o}, Q_I]$ *(resp., type I truncated wedge* $\overline{\mathbf{W}}_I = [\mathbf{o}, \overline{Q}_I]$*). Similarly, we define the* $(d-1)$*-dimensional bases* Q_{II} *and* $\overline{Q}_{II}$ *of* $\mathbf{W}_{II}$ *and* $\overline{\mathbf{W}}_{II}$*. Finally, let*

$$h_1 = \|\mathbf{r}_1\|, h_2 = \|\mathbf{r}_2 - \mathbf{r}_1\|, \ldots, h_{d-2} = \|\mathbf{r}_{d-2} - \mathbf{r}_{d-3}\|.$$

Lemma 7.3.15 *Let* $\mathbf{W}_I$ *(resp.,* $\mathbf{W}_{II}$*) denote the wedge of type I (resp., of type II) with the 2-dimensional base* F_2 *which is generated by the* $(d-2)$*-dimensional Rogers orthoscheme* $\mathrm{conv}\{\mathbf{o}, \mathbf{r}_1, \ldots, \mathbf{r}_{d-2}\}$ *of the Voronoi polytope* $\mathbf{P} \subset \mathbb{E}^d, d \geq 4$*. Then we have the following volume formulas.*

(1) $\mathrm{vol}_{d-1}(Q_I) = \frac{2}{(d-1)!} \left(\prod_{i=2}^{d-2} h_i \right) \mathrm{vol}_2(F_2)$ *and*

(2) $\mathrm{vol}_d(\mathbf{W}_I) = \frac{2}{d!} \left(\prod_{i=1}^{d-2} h_i \right) \mathrm{vol}_2(F_2)$*.*

  Similar formulas hold for the corresponding dimensional volumes of $\overline{Q}_I$*,* $\overline{\mathbf{W}}_I$*,* Q_{II}*,* $\mathbf{W}_{II}$*,* $\overline{Q}_{II}$*, and* $\overline{\mathbf{W}}_{II}$*.*

  In general, if $K \subset \mathrm{aff} F_2$ *is a convex domain, then*

(3) $\mathrm{vol}_{d-1}([\mathbf{r}_1, \ldots, \mathbf{r}_{d-3}, K]) = \frac{2}{(d-1)!} \left(\prod_{i=2}^{d-2} h_i \right) \mathrm{vol}_2(K)$ *and*

(4) $\mathrm{vol}_d([\mathbf{o}, \mathbf{r}_1, \ldots, \mathbf{r}_{d-3}, K]) = \frac{2}{d!} \left(\prod_{i=1}^{d-2} h_i \right) \mathrm{vol}_2(K)$*.*

Proof: The proof follows from Lemma 7.3.6 and Lemma 7.3.9 in a straightforward way. $\square$

7.3.5 The integral representation of surface density in (truncated) wedges

The central notion of this section is the limiting surface density introduced as follows.

Definition 7.3.16 *Let* $\mathbf{W}_I$ *(resp.,* $\mathbf{W}_{II}$*) denote the wedge of type I (resp., of type II) with the 2-dimensional base* F_2 *which is generated by the* $(d-2)$*-dimensional Rogers orthoscheme* $\mathrm{conv}\{\mathbf{o}, \mathbf{r}_1, \ldots, \mathbf{r}_{d-2}\}$ *of the Voronoi polytope* $\mathbf{P} \subset \mathbb{E}^d, d \geq 4$*. Then choose a coordinate system with two perpendicular axes in the plane* $\mathrm{aff} F_2$ *meeting at the point* $\mathbf{r}_{d-2}$*. Now, if* $\mathbf{x}$ *is an arbitrary point of the plane* $\mathrm{aff} F_2$*, then for a positive integer* n *let* $T_n(\mathbf{x}) \subset \mathrm{aff} F_2$ *denote the square centered at* $\mathbf{x}$ *having sides of length* $\frac{1}{n}$ *parallel to the fixed coordinate axes. Then the limiting surface density* $\widehat{\delta}_{\lim}([\mathbf{o}, \mathbf{r}_1, \ldots, \mathbf{r}_{d-3}, \mathbf{x}], S)$ *of the* $(d-1)$*-dimensional unit sphere* S *in the* $(d-2)$*-dimensional orthoscheme* $[\mathbf{o}, \mathbf{r}_1, \ldots, \mathbf{r}_{d-3}, \mathbf{x}]$ *is defined by*

$$\widehat{\delta}_{\mathrm{lim}}\left(\left[\mathbf{o},\mathbf{r}_1,\ldots,\mathbf{r}_{d-3},\mathbf{x}\right],S\right) = \lim_{n\to\infty}\widehat{\delta}\left(\left[\mathbf{o},\mathbf{r}_1,\ldots,\mathbf{r}_{d-3},T_n(\mathbf{x})\right],S\right).$$

Based on this we are able to give an integral representation of the surface density in a (truncated) wedge.

Lemma 7.3.17 *Let $\mathbf{W}_I$ (resp., $\mathbf{W}_{II}$) denote the wedge of type I (resp., of type II) with the 2-dimensional base F_2 which is generated by the $(d-2)$-dimensional Rogers orthoscheme $\mathrm{conv}\{\mathbf{o},\mathbf{r}_1,\ldots,\mathbf{r}_{d-2}\}$ of the Voronoi polytope $\mathbf{P}\subset\mathbb{E}^d, d\geq 4$.*
(1) If $\mathbf{x}\in\mathrm{aff}\,F_2$ and $\mathbf{y}\in\mathrm{aff}\,F_2$ are points such that $\|\mathbf{x}\|\leq\|\mathbf{y}\|$, then

$$\widehat{\delta}_{\mathrm{lim}}\left(\left[\mathbf{o},\mathbf{r}_1,\ldots,\mathbf{r}_{d-3},\mathbf{x}\right],S\right)\geq\widehat{\delta}_{\mathrm{lim}}\left(\left[\mathbf{o},\mathbf{r}_1,\ldots,\mathbf{r}_{d-3},\mathbf{y}\right],S\right).$$

(2) For the surface densities of the unit sphere S in the wedge $\mathbf{W}_I$ and in the truncated wedge $\overline{\mathbf{W}}_I$ we have the following formulas.

$$\widehat{\delta}(\mathbf{W}_I,S) = \frac{\mathrm{Svol}_{d-1}([\mathbf{o},Q_I]\cap S)}{\mathrm{vol}_{d-1}(Q_I)}$$

$$= \frac{1}{\mathrm{vol}_2(F_2)}\int_{F_2}\widehat{\delta}_{\mathrm{lim}}\left(\left[\mathbf{o},\mathbf{r}_1,\ldots,\mathbf{r}_{d-3},\mathbf{x}\right],S\right)\ dx$$

and

$$\widehat{\delta}(\overline{\mathbf{W}}_I,S) = \frac{\mathrm{Svol}_{d-1}([\mathbf{o},\overline{Q}_I]\cap S)}{\mathrm{vol}_{d-1}(\overline{Q}_I)}$$

$$= \frac{1}{\mathrm{vol}_2(\overline{\mathbf{B}}\cap F_2)}\int_{\overline{\mathbf{B}}\cap F_2}\widehat{\delta}_{\mathrm{lim}}\left(\left[\mathbf{o},\mathbf{r}_1,\ldots,\mathbf{r}_{d-3},\mathbf{x}\right],S\right)\ dx,$$

where dx stands for the Euclidean area element in the plane $\mathrm{aff}\,F_2$. Similar formulas hold for $\mathbf{W}_{II}$ and $\overline{\mathbf{W}}_{II}$.
(3) In general, if $K\subset\mathrm{aff}\,F_2$ is a convex domain, then the surface density of the unit sphere S in the d-dimensional convex cone $[\mathbf{o},\mathbf{r}_1,\ldots,\mathbf{r}_{d-3},K]$ with apex $\mathbf{o}$ and $(d-1)$-dimensional base $[\mathbf{r}_1,\ldots,\mathbf{r}_{d-3},K]$ can be computed as follows.

$$\widehat{\delta}([\mathbf{o},\mathbf{r}_1,\ldots,\mathbf{r}_{d-3},K],S) = \frac{1}{\mathrm{vol}_2(K)}\int_K\widehat{\delta}_{\mathrm{lim}}\left(\left[\mathbf{o},\mathbf{r}_1,\ldots,\mathbf{r}_{d-3},\mathbf{x}\right],S\right)\ dx.$$

Proof:
(1) It is sufficient to look at the case $\|\mathbf{x}\| < \|\mathbf{y}\|$. (The case $\|\mathbf{x}\| = \|\mathbf{y}\|$ follows from this by standard limit procedure.) Then recall that

$$\widehat{\delta}\left(\left[\mathbf{o},\mathbf{r}_1,\ldots,\mathbf{r}_{d-3},T_n(\mathbf{x})\right],S\right) = h_1\delta\left(\left[\mathbf{o},\mathbf{r}_1,\ldots,\mathbf{r}_{d-3},T_n(\mathbf{x})\right],S\right)$$

and

$$\widehat{\delta}\left(\left[\mathbf{o},\mathbf{r}_1,\ldots,\mathbf{r}_{d-3},T_n(\mathbf{y})\right],S\right) = h_1\delta\left(\left[\mathbf{o},\mathbf{r}_1,\ldots,\mathbf{r}_{d-3},T_n(\mathbf{y})\right],S\right).$$

Thus, it is sufficient to show that if n is sufficiently large, then

$$\delta\left([\mathbf{o},\mathbf{r}_1,\ldots,\mathbf{r}_{d-3},T_n(\mathbf{x})],S\right) \geq \delta\left([\mathbf{o},\mathbf{r}_1,\ldots,\mathbf{r}_{d-3},T_n(\mathbf{y})],S\right).$$

This we can get as follows. We can approximate the d-dimensional convex cone $[\mathbf{o},\mathbf{r}_1,\ldots,\mathbf{r}_{d-3},T_n(\mathbf{x})]$ (resp., $[\mathbf{o},\mathbf{r}_1,\ldots,\mathbf{r}_{d-3},T_n(\mathbf{y})]$) arbitrarily close with a finite (but possibly large) number of non-overlapping d-dimensional orthoschemes each containing the $(d-3)$-dimensional orthoscheme $[\mathbf{o},\mathbf{r}_1,\ldots,\mathbf{r}_{d-3}]$ as a face and each having all the edge lengths of the 3 edges going out from the vertex $\mathbf{o}$ and not lying on the face $[\mathbf{o},\mathbf{r}_1,\ldots,\mathbf{r}_{d-3}]$ close to $\|\mathbf{x}\|$ (resp., $\|\mathbf{y}\|$) for n sufficiently large (see also Lemma 7.3.9). Thus, the claim follows from (1) of Lemma 7.3.11 rather easily.

(2),(3) It is sufficient to prove the corresponding formula for K.

A typical term of the Riemann–Lebesgue sum of

$$\frac{1}{\mathrm{vol}_2(K)}\int_K \widehat{\delta}_{\lim}\left([\mathbf{o},\mathbf{r}_1,\ldots,\mathbf{r}_{d-3},\mathbf{x}],S\right)\,d\mathbf{x}$$

is equal to

$$\frac{1}{\mathrm{vol}_2(K)}\widehat{\delta}\left([\mathbf{o},\mathbf{r}_1,\ldots,\mathbf{r}_{d-3},T_n(\mathbf{x}_m)],S\right)\mathrm{vol}_2(T_n(\mathbf{x}_m)), m \in M.$$

Using Lemma 7.3.15 this turns out to be equal to

$$\frac{\mathrm{vol}_{d-1}([\mathbf{r}_1,\ldots,\mathbf{r}_{d-3},T_n(\mathbf{x}_m)])}{\mathrm{vol}_{d-1}([\mathbf{r}_1,\ldots,\mathbf{r}_{d-3},K])}\widehat{\delta}\left([\mathbf{o},\mathbf{r}_1,\ldots,\mathbf{r}_{d-3},T_n(\mathbf{x}_m)],S\right)$$

$$=\frac{\mathrm{Svol}_{d-1}([\mathbf{o},\mathbf{r}_1,\ldots,\mathbf{r}_{d-3},T_n(\mathbf{x}_m)]\cap S)}{\mathrm{vol}_{d-1}([\mathbf{r}_1,\ldots,\mathbf{r}_{d-3},K])}.$$

Finally, as the union of the non-overlapping squares $T_n(\mathbf{x}_m), m \in M$ is a good approximation of the convex domain K in the plane $\mathrm{aff}\,F_2$ we get that

$$\sum_{m\in M}\frac{\mathrm{Svol}_{d-1}([\mathbf{o},\mathbf{r}_1,\ldots,\mathbf{r}_{d-3},T_n(\mathbf{x}_m)]\cap S)}{\mathrm{vol}_{d-1}([\mathbf{r}_1,\ldots,\mathbf{r}_{d-3},K])}$$

$$=\frac{\sum_{m\in M}\mathrm{Svol}_{d-1}([\mathbf{o},\mathbf{r}_1,\ldots,\mathbf{r}_{d-3},T_n(\mathbf{x}_m)]\cap S)}{\mathrm{vol}_{d-1}([\mathbf{r}_1,\ldots,\mathbf{r}_{d-3},K])}$$

is a good approximation of

$$\frac{\mathrm{Svol}_{d-1}\left([\mathbf{o},\mathbf{r}_1,\ldots,\mathbf{r}_{d-3},K]\cap S\right)}{\mathrm{vol}_{d-1}([\mathbf{r}_1,\ldots,\mathbf{r}_{d-3},K])}=\widehat{\delta}([\mathbf{o},\mathbf{r}_1,\ldots,\mathbf{r}_{d-3},K],S).$$

This completes the proof of Lemma 7.3.17.   □

7.3.6 Truncation of wedges increases the surface density

Lemma 7.3.18 *Let $\mathbf{W}_I$ (resp., $\mathbf{W}_{II}$) denote the wedge of type I (resp., of type II) with the 2-dimensional base F_2 which is generated by the $(d-2)$-dimensional Rogers orthoscheme $\mathrm{conv}\{\mathbf{o}, \mathbf{r}_1, \ldots, \mathbf{r}_{d-2}\}$ of the Voronoi polytope $\mathbf{P} \subset \mathbb{E}^d, d \geq 4$. Then*

$$\widehat{\delta}(\mathbf{W}_I, S) \leq \widehat{\delta}(\overline{\mathbf{W}}_I, S) \quad \left(\text{resp.,}\ \widehat{\delta}(\mathbf{W}_{II}, S) \leq \widehat{\delta}(\overline{\mathbf{W}}_{II}, S)\right).$$

Proof: Notice that (1) of Lemma 7.3.17 easily implies that if $0 < \mathrm{vol}_2(F_2 \setminus \overline{\mathbf{B}})$, then for any $\mathbf{x}^* \in F_2$ with $\|\mathbf{x}^*\| = \sqrt{\frac{2d}{d+1}}$ we have that

$$\frac{1}{\mathrm{vol}_2(F_2 \setminus \overline{\mathbf{B}})} \int_{F_2 \setminus \overline{\mathbf{B}}} \widehat{\delta}_{\lim}\left([\mathbf{o}, \mathbf{r}_1, \ldots, \mathbf{r}_{d-3}, \mathbf{x}], S\right) d\mathbf{x}$$

$$\leq \widehat{\delta}_{\lim}\left([\mathbf{o}, \mathbf{r}_1, \ldots, \mathbf{r}_{d-3}, \mathbf{x}^*], S\right)$$

$$\leq \frac{1}{\mathrm{vol}_2(\overline{\mathbf{B}} \cap F_2)} \int_{\overline{\mathbf{B}} \cap F_2} \widehat{\delta}_{\lim}\left([\mathbf{o}, \mathbf{r}_1, \ldots, \mathbf{r}_{d-3}, \mathbf{x}], S\right) d\mathbf{x}.$$

Thus, if $0 < \mathrm{vol}_2(F_2 \setminus \overline{\mathbf{B}})$, then (2) of Lemma 7.3.17 yields that

$$\widehat{\delta}(\mathbf{W}_I, S) = \frac{1}{\mathrm{vol}_2(F_2)} \int_{F_2} \widehat{\delta}_{\lim}\left([\mathbf{o}, \mathbf{r}_1, \ldots, \mathbf{r}_{d-3}, \mathbf{x}], S\right) d\mathbf{x}$$

$$= \frac{\mathrm{vol}_2(\overline{\mathbf{B}} \cap F_2)}{\mathrm{vol}_2(F_2)} \cdot \frac{1}{\mathrm{vol}_2(\overline{\mathbf{B}} \cap F_2)} \int_{\overline{\mathbf{B}} \cap F_2} \widehat{\delta}_{\lim}\left([\mathbf{o}, \mathbf{r}_1, \ldots, \mathbf{r}_{d-3}, \mathbf{x}], S\right) d\mathbf{x}$$

$$+ \frac{\mathrm{vol}_2(F_2 \setminus \overline{\mathbf{B}})}{\mathrm{vol}_2(F_2)} \cdot \frac{1}{\mathrm{vol}_2(F_2 \setminus \overline{\mathbf{B}})} \int_{F_2 \setminus \overline{\mathbf{B}}} \widehat{\delta}_{\lim}\left([\mathbf{o}, \mathbf{r}_1, \ldots, \mathbf{r}_{d-3}, \mathbf{x}], S\right) d\mathbf{x}$$

$$\leq \frac{\mathrm{vol}_2(\overline{\mathbf{B}} \cap F_2)}{\mathrm{vol}_2(F_2)} \cdot \frac{1}{\mathrm{vol}_2(\overline{\mathbf{B}} \cap F_2)} \int_{\overline{\mathbf{B}} \cap F_2} \widehat{\delta}_{\lim}\left([\mathbf{o}, \mathbf{r}_1, \ldots, \mathbf{r}_{d-3}, \mathbf{x}], S\right) d\mathbf{x}$$

$$+ \frac{\mathrm{vol}_2(F_2 \setminus \overline{\mathbf{B}})}{\mathrm{vol}_2(F_2)} \cdot \frac{1}{\mathrm{vol}_2(\overline{\mathbf{B}} \cap F_2)} \int_{\overline{\mathbf{B}} \cap F_2} \widehat{\delta}_{\lim}\left([\mathbf{o}, \mathbf{r}_1, \ldots, \mathbf{r}_{d-3}, \mathbf{x}], S\right) d\mathbf{x}$$

$$= \frac{1}{\mathrm{vol}_2(\overline{\mathbf{B}} \cap F_2)} \int_{\overline{\mathbf{B}} \cap F_2} \widehat{\delta}_{\lim}\left([\mathbf{o}, \mathbf{r}_1, \ldots, \mathbf{r}_{d-3}, \mathbf{x}], S\right) d\mathbf{x} = \widehat{\delta}(\overline{\mathbf{W}}_I, S).$$

As the same method works for $\mathbf{W}_{II}$ and $\overline{\mathbf{W}}_{II}$ this completes the proof of Lemma 7.3.18. □

7.3.7 Maximum surface density in truncated wedges of type I

Let $\overline{\mathbf{W}}_I$ denote the truncated wedge of type I with the 2-dimensional base $\overline{\mathbf{B}} \cap F_2$ which is generated by the $(d-2)$-dimensional Rogers orthoscheme $\mathrm{conv}\{\mathbf{o}, \mathbf{r}_1, \ldots, \mathbf{r}_{d-2}\}$ of the Voronoi polytope $\mathbf{P} \subset \mathbb{E}^d, d \geq 8$. By assumption F_2 is a 2-dimensional face of the Voronoi polytope $\mathbf{P}$ with

$$\sqrt{\frac{2(d-2)}{d-1}} \leq h = R(F_2) < \sqrt{\frac{2(d-1)}{d}}.$$

Let $G_0 \subset \mathrm{aff}\, F_2$ (resp., $G \subset \mathrm{aff}\, F_2$) denote the closed circular disk of radius

$$g_0(h) = \sqrt{\frac{2d}{d+1} - h^2} \left(\text{resp.,}\ g(h) = \frac{2-h^2}{\sqrt{4-h^2}} \right)$$

centered at the point $\mathbf{r}_{d-2}$. It is easy to see that $G \subset \mathrm{relint}\, G_0$ for all $\sqrt{\frac{2(d-2)}{d-1}} \leq h < \sqrt{\frac{2(d-1)}{d}}$. (Moreover $G = G_0$ for $h = \sqrt{\frac{2(d-1)}{d}}$.) Notice that $G_0 = \overline{\mathbf{B}} \cap \mathrm{aff}\, F_2$, thus Corollary 7.3.2 implies that there is no vertex of the face F_2 belonging to the relative interior of G_0. Moreover, as $h = R(F_2) < \sqrt{2}$ Lemma 7.3.1 yields that $\frac{2}{\sqrt{4-h^2}} \leq R(F_1)$ holds for any side F_1 of the face F_2, hence $G \subset F_2$ and of course, $G \subset \overline{\mathbf{B}} \cap F_2 = G_0 \cap F_2$. Now, let $M \subset \mathrm{aff}\, F_2$ be a square circumscribed about G. A straightforward computation yields that $\frac{g_0(h)}{g(h)}$ is a strictly decreasing function on the interval $\left[\sqrt{\frac{2(d-2)}{d-1}}, \sqrt{\frac{2(d-1)}{d}} \right)$ (i.e., $\frac{d}{dh}\left(\frac{g_0(h)}{g(h)}\right) < 0$ on the interval $\left(\sqrt{\frac{2(d-2)}{d-1}}, \sqrt{\frac{2(d-1)}{d}}\right)$) and

$$\frac{g_0\left(\sqrt{\frac{2(d-2)}{d-1}}\right)}{g\left(\sqrt{\frac{2(d-2)}{d-1}}\right)} = \sqrt{\frac{2d}{d+1}} < \sqrt{2}.$$

Thus, the vertices of the square M do not belong to G_0. Finally, as $d \geq 8$ Lemma 7.3.10 implies that there are at most four sides of the face F_2 that intersect the relative interior of G_0.

The following statement is rather natural from the point of view of the local geometry introduced above, however, its three-page proof based on Lemma 7.3.11 and Lemma 7.3.17 published in [56] is a bit technical and so, for that reason we do not prove it here; instead we refer the interested reader to the proper section in [56].

Lemma 7.3.19 *Let* $\overline{\mathbf{W}}_I$ *denote the truncated wedge of type I with the 2-dimensional base* $\overline{\mathbf{B}} \cap F_2$ *which is generated by the* $(d-2)$-*dimensional Rogers orthoscheme* $\mathrm{conv}\{\mathbf{o}, \mathbf{r}_1, \ldots, \mathbf{r}_{d-2}\}$ *of the Voronoi polytope* $\mathbf{P} \subset \mathbb{E}^d, d \geq 8$. *Then*

$$\widehat{\delta}(\overline{\mathbf{W}}_I, S) \leq \widehat{\delta}([\mathbf{o}, \mathbf{r}_1, \ldots, \mathbf{r}_{d-3}, G_0 \cap M], S).$$

It is clear from the construction that we can write $\widehat{\delta}([\mathbf{o}, \mathbf{r}_1, \ldots, \mathbf{r}_{d-3}, G_0 \cap M], S)$ as a function of $d-2$ variables, namely

$$\widehat{\Delta}(\xi_1, \ldots, \xi_{d-3}, \xi_{d-2}) = \widehat{\delta}([\mathbf{o}, \mathbf{r}_1, \ldots, \mathbf{r}_{d-3}, G_0 \cap M], S),$$

where $\xi_1 = \|\mathbf{r}_1\|, \ldots, \xi_{d-3} = \|\mathbf{r}_{d-3}\|, \xi_{d-2} = \|\mathbf{r}_{d-2}\| = h$. Corollary 7.3.2 and the assumption on h imply that

$$m_1 = 1 \le \xi_1, \ldots, m_i = \sqrt{\frac{2i}{i+1}} \le \xi_i, \ldots, m_{d-3} = \sqrt{\frac{2(d-3)}{d-2}} \le \xi_{d-3},$$

$$m_{d-2} = \sqrt{\frac{2(d-2)}{d-1}} \le \xi_{d-2} = h < \sqrt{\frac{2(d-1)}{d}}.$$

Notice that if $\|\mathbf{r}_i\| = m_i$ for all $1 \le i \le d-2$, then $[\mathbf{o}, \mathbf{r}_1, \ldots, \mathbf{r}_{d-3}, G_0 \cap M]$ can be dissected into four pieces each being congruent to $\mathbf{W}$ and therefore $\widehat{\delta}([\mathbf{o}, \mathbf{r}_1, \ldots, \mathbf{r}_{d-3}, G_0 \cap M], S) = \widehat{\sigma}_d$.

Lemma 7.3.20

$$\widehat{\Delta}(\xi_1, \ldots, \xi_{d-3}, \xi_{d-2}) \le \widehat{\Delta}(m_1, \ldots, m_{d-3}, m_{d-2}) = \widehat{\sigma}_d.$$

Proof: For any fixed $\xi_{d-2} = h$, (2) of Lemma 7.3.11 easily implies that

$$\widehat{\Delta}(\xi_1, \ldots, \xi_{d-3}, h) \le \widehat{\Delta}(m_1, \ldots, m_{d-3}, h).$$

Finally, using Lemma 7.3.11 again, it is rather straightforward to show that the function $\widehat{\Delta}(m_1, \ldots, m_{d-3}, h)$ as a function of h is decreasing on the interval $\left(\sqrt{\frac{2(d-2)}{d-1}}, \sqrt{\frac{2(d-1)}{d}}\right)$. From this it follows that

$$\widehat{\Delta}(m_1, \ldots, m_{d-3}, h) \le \widehat{\Delta}(m_1, \ldots, m_{d-3}, m_{d-2}) = \widehat{\sigma}_d,$$

finishing the proof of Lemma 7.3.20. □

Thus, Lemma 7.3.19 and Lemma 7.3.20 yield the following immediate estimate.

Corollary 7.3.21 *Let $\overline{\mathbf{W}}_I$ denote the truncated wedge of type I with the 2-dimensional base $\overline{\mathbf{B}} \cap F_2$ which is generated by the $(d-2)$-dimensional Rogers orthoscheme $\mathrm{conv}\{\mathbf{o}, \mathbf{r}_1, \ldots, \mathbf{r}_{d-2}\}$ of the Voronoi polytope $\mathbf{P} \subset \mathbb{E}^d, d \ge 8$. Then*

$$\widehat{\delta}(\overline{\mathbf{W}}_I, S) \le \widehat{\sigma}_d.$$

7.3.8 An upper bound for the surface density in truncated wedges of type II

It is sufficient to prove the following statement.

Lemma 7.3.22 *Let $\overline{\mathbf{W}}_{II}$ denote the truncated wedge of type II with the 2-dimensional base $\overline{\mathbf{B}} \cap F_2$ which is generated by the $(d-2)$-dimensional Rogers orthoscheme $\mathrm{conv}\{\mathbf{o}, \mathbf{r}_1, \ldots, \mathbf{r}_{d-2}\}$ of the Voronoi polytope $\mathbf{P} \subset \mathbb{E}^d, d \geq 4$. Then*

$$\widehat{\delta}(\overline{\mathbf{W}}_{II}, S) \leq \widehat{\sigma}_d.$$

Proof: By assumption F_2 is a 2-dimensional face of the Voronoi polytope $\mathbf{P}$ with

$$\sqrt{\frac{2(d-1)}{d}} \leq h = R(F_2) < \sqrt{\frac{2d}{d+1}}.$$

Let $G_0 \subset \mathrm{aff}\, F_2$ denote the closed circular disk of radius $g_0(h) = \sqrt{\frac{2d}{d+1} - h^2}$ centered at the point $\mathbf{r}_{d-2}$. As $h = R(F_2) < \sqrt{2}$, therefore Lemma 7.3.1 yields that

$$\sqrt{\frac{2d}{d+1}} \leq \frac{2}{\sqrt{4-h^2}} \leq R(F_1)$$

holds for any side F_1 of the face F_2. Thus,

$$\overline{\mathbf{B}} \cap F_2 = G_0$$

and so

$$\widehat{\delta}(\overline{\mathbf{W}}_{II}, S) = \widehat{\delta}([\mathbf{o}, \mathbf{r}_1, \ldots, \mathbf{r}_{d-3}, G_0], S).$$

It is clear from the construction that we can write $\widehat{\delta}([\mathbf{o}, \mathbf{r}_1, \ldots, \mathbf{r}_{d-3}, G_0], S)$ as a function of $d-2$ variables, namely

$$\widehat{\Delta}^*(\xi_1, \ldots, \xi_{d-3}, \xi_{d-2}) = \widehat{\delta}([\mathbf{o}, \mathbf{r}_1, \ldots, \mathbf{r}_{d-3}, G_0], S),$$

where $\xi_1 = \|\mathbf{r}_1\|, \ldots, \xi_{d-3} = \|\mathbf{r}_{d-3}\|, \xi_{d-2} = \|\mathbf{r}_{d-2}\| = h$. Corollary 7.3.2 and the assumption on h imply that

$$m_1 = 1 \leq \xi_1, \ldots, m_i = \sqrt{\frac{2i}{i+1}} \leq \xi_i, \ldots, m_{d-3} = \sqrt{\frac{2(d-3)}{d-2}} \leq \xi_{d-3},$$

$$m_{d-2}^* = \sqrt{\frac{2(d-1)}{d}} \leq \xi_{d-2} = h < \sqrt{\frac{2d}{d+1}}.$$

For any fixed $\xi_{d-2} = h$, (2) of Lemma 7.3.11 easily implies that

$$\widehat{\Delta}^*(\xi_1, \ldots, \xi_{d-3}, h) \leq \widehat{\Delta}^*(m_1, \ldots, m_{d-3}, h).$$

Finally, again applying (2) of Lemma 7.3.11 we immediately get that

$$\widehat{\Delta}^*(m_1, \ldots, m_{d-3}, h) \leq \widehat{\Delta}^*(m_1, \ldots, m_{d-3}, m_{d-2}^*) \leq \widehat{\sigma}_d.$$

This completes the proof of Lemma 7.3.22. □

7.3.9 The overall estimate of surface density in Voronoi cells

Let $\mathbf{P}$ be a d-dimensional Voronoi polytope of a packing $\mathcal{P}$ of d-dimensional unit balls in $\mathbb{E}^d, d \geq 8$. Without loss of generality we may assume that the unit ball $\mathbf{B} = \{\mathbf{x} \in \mathbb{E}^d \mid \text{dist}(\mathbf{o}, \mathbf{x}) = \|\mathbf{x}\| \leq 1\}$ centered at the origin $\mathbf{o}$ of $\mathbb{E}^d$ is one of the unit balls of $\mathcal{P}$ with $\mathbf{P}$ as its Voronoi cell. As before, let S denote the boundary of $\mathbf{B}$.

First, we dissect $\mathbf{P}$ into d-dimensional Rogers simplices. Then let $\text{conv}\{\mathbf{o}, \mathbf{r}_1, \ldots, \mathbf{r}_d\}$ be one of these d-dimensional Rogers simplices assigned to the flag say, $F_0 \subset \cdots \subset F_{d-1}$ of $\mathbf{P}$. As $\mathbf{r}_i \in F_{d-i}, 1 \leq i \leq d$ it is clear that $\text{aff}\{\mathbf{r}_{d-2}, \mathbf{r}_{d-1}, \mathbf{r}_d\} = \text{aff} F_2$ and so

$$\text{dist}(\mathbf{o}, \text{aff}\{\mathbf{r}_{d-2}, \mathbf{r}_{d-1}, \mathbf{r}_d\}) = \text{dist}(\mathbf{o}, \text{aff} F_2) = R(F_2).$$

Notice that Corollary 7.3.2 implies that $\sqrt{\frac{2(d-2)}{d-1}} \leq R(F_2)$.

Second, we group the d-dimensional Rogers simplices of $\mathbf{P}$ as follows.

(1): If $\sqrt{\frac{2(d-2)}{d-1}} \leq R(F_2) < \sqrt{\frac{2(d-1)}{d}}$, then we assign the Rogers simplex $\text{conv}\{\mathbf{o}, \mathbf{r}_1, \ldots, \mathbf{r}_d\}$ to the type I wedge $\mathbf{W}_I$ with the 2-dimensional base F_2 generated by the $(d-2)$-dimensional Rogers orthoscheme $\text{conv}\{\mathbf{o}, \mathbf{r}_1, \ldots, \mathbf{r}_{d-2}\}$ of the Voronoi polytope $\mathbf{P} \subset \mathbb{E}^d, d \geq 8$.

(2): If $\sqrt{\frac{2(d-1)}{d}} \leq R(F_2) < \sqrt{\frac{2d}{d+1}}$, then we assign the Rogers simplex $\text{conv}\{\mathbf{o}, \mathbf{r}_1, \ldots, \mathbf{r}_d\}$ to the type II wedge $\mathbf{W}_{II}$ with the 2-dimensional base F_2 generated by the $(d-2)$-dimensional Rogers orthoscheme $\text{conv}\{\mathbf{o}, \mathbf{r}_1, \ldots, \mathbf{r}_{d-2}\}$ of the Voronoi polytope $\mathbf{P} \subset \mathbb{E}^d, d \geq 8$.

(3): If $\sqrt{\frac{2d}{d+1}} \leq R(F_2)$, then we assign the Rogers simplex $\text{conv}\{\mathbf{o}, \mathbf{r}_1, \ldots, \mathbf{r}_d\}$ to itself as the type III wedge $\mathbf{W}_{III}$.

As the wedges of types I, II, and III of the given Voronoi polytope $\mathbf{P}$ sit over the 2-skeleton of $\mathbf{P}$ and form a tiling of $\mathbf{P}$ it is clear that each d-dimensional Rogers simplex of $\mathbf{P}$ belongs to exactly one of them. As a result, in order to show that the surface density $\widehat{\delta}(\mathbf{P}, S) = \frac{\text{Svol}_{d-1}(S)}{\text{svol}_{d-1}(\text{bd}\mathbf{P})} = \frac{d\omega_d}{\text{svol}_{d-1}(\text{bd}\mathbf{P})}$ of the unit sphere S in the Voronoi polytope $\mathbf{P}$ is bounded from above by $\widehat{\sigma}_d$, it is sufficient to prove the following inequalities.

$\widehat{(1)}$: $\widehat{\delta}(W_I, S) \leq \widehat{\sigma}_d$,

$\widehat{(2)}$: $\widehat{\delta}(W_{II}, S) \leq \widehat{\sigma}_d$,

$\widehat{(3)}$: $\widehat{\delta}(W_{III}, S) \leq \widehat{\sigma}_d$.

This final task is now easy. Namely, Lemma 7.3.18, Corollary 7.3.21, and Lemma 7.3.22 yield $\widehat{(1)}$ and $\widehat{(2)}$ in a straightforward way. Finally, $\widehat{(3)}$ follows with the help of (2) of Lemma 7.3.11 rather easily.

For the details of the proof of $\widehat{\sigma}_d < \sigma_d$, based on the so-called "Lemma of Strict Comparison", we refer the interested reader to the proper section in [56].

This completes the proof of Theorem 1.4.8.

7.4 Proof of Theorem 1.7.3

7.4.1 The signed volume of convex polytopes

Definition 7.4.1 *Let* $\mathbf{P} := \mathrm{conv}\{\mathbf{p}_1, \mathbf{p}_2, \ldots, \mathbf{p}_n\}$ *be a d-dimensional convex polytope in* $\mathbb{E}^d, d \geq 2$ *with vertices* $\mathbf{p}_1, \mathbf{p}_2, \ldots, \mathbf{p}_n$. *If* $F := \mathrm{conv}\{\mathbf{p}_{i_1}, \ldots, \mathbf{p}_{i_k}\}$ *is an arbitrary face of* $\mathbf{P}$, *then the barycenter of* F *is*

$$\mathbf{c}_F := \frac{1}{k} \sum_{j=1}^{k} \mathbf{p}_{i_j}. \tag{7.10}$$

Let $F_0 \subset F_1 \subset \cdots \subset F_l, 0 \leq l \leq d-1$ denote a sequence of faces, called a (partial) flag of $\mathbf{P}$, where F_0 is a vertex and F_{i-1} is a facet (a face one dimension lower) of F_i for $i = 1, \ldots, l$. Then the simplices of the form $\mathrm{conv}\{\mathbf{c}_{F_0}, \mathbf{c}_{F_1}, \ldots, \mathbf{c}_{F_l}\}$ constitute a simplicial complex $\mathcal{C}_{\mathbf{P}}$ whose underlying space is the boundary of $\mathbf{P}$.

We regard all points in $\mathbb{E}^d$ as row vectors and use $\mathbf{q}^T$ for the column vector that is the transpose of the row vector $\mathbf{q}$. Moreover, $[\mathbf{q}_1, \ldots, \mathbf{q}_d]$ is the (square) matrix with the ith row $\mathbf{q}_i$.

Choosing a $(d-1)$-dimensional simplex of $\mathcal{C}_{\mathbf{P}}$ to be positively oriented, one can check whether the orientation of an arbitrary $(d-1)$-dimensional simplex $\mathrm{conv}\{\mathbf{c}_{F_0}, \mathbf{c}_{F_1}, \ldots, \mathbf{c}_{F_{d-1}}\}$ of $\mathcal{C}_{\mathbf{P}}$ (generated by the given sequence of its vertices), is positive or negative. Let $\mathrm{sign}\left(\mathrm{conv}\{\mathbf{c}_{F_0}, \mathbf{c}_{F_1}, \ldots, \mathbf{c}_{F_{d-1}}\}\right)$ be equal to 1 (resp., -1) if the orientation of the $(d-1)$-dimensional simplex $\mathrm{conv}\{\mathbf{c}_{F_0}, \mathbf{c}_{F_1}, \ldots, \mathbf{c}_{F_{d-1}}\}$ is positive (resp., negative).

Definition 7.4.2 *The signed volume* $V(\mathbf{P})$ *of* $\mathbf{P}$ *is defined as*

$$\frac{1}{d!} \sum_{F_0 \subset F_1 \subset \cdots \subset F_{d-1}} \mathrm{sign}\left(\mathrm{conv}\{\mathbf{c}_{F_0}, \mathbf{c}_{F_1}, \ldots, \mathbf{c}_{F_{d-1}}\}\right) \det[\mathbf{c}_{F_0}, \mathbf{c}_{F_1}, \ldots, \mathbf{c}_{F_{d-1}}],$$

$$\tag{7.11}$$

where the sum is taken over all flags of faces $F_0 \subset F_1 \subset \cdots \subset F_{d-1}$ *of* $\mathbf{P}$, *and* $\det[\cdot]$ *is the determinant function.*

The following is clear.

Lemma 7.4.3

$$V(\mathbf{P}) = \frac{1}{d!} \sum_{F_0 \subset \cdots \subset F_{d-1}} \mathrm{sign}\left(\mathrm{conv}\{\mathbf{c}_{F_0}, \mathbf{c}_{F_1}, \ldots, \mathbf{c}_{F_{d-1}}\}\right) \mathbf{c}_{F_0} \wedge \mathbf{c}_{F_1} \wedge \cdots \wedge \mathbf{c}_{F_{d-1}},$$

where $\wedge$ *stands for the wedge product of vectors. Moreover, one can choose the orientation of the boundary of* $\mathbf{P}$ *such that* $V(\mathbf{P}) = \mathrm{vol}_d(\mathbf{P})$, *where* $\mathrm{vol}_d(\cdot)$ *refers to the d-dimensional volume measure in* $\mathbb{E}^d, d \geq 2$.

7.4.2 The volume force of convex polytopes

We wish to compute the gradient of $V(\mathbf{P})$, where $\mathbf{P} = \mathrm{conv}\{\mathbf{p}_1, \mathbf{p}_2, \ldots, \mathbf{p}_n\}$ is regarded as a function of its vertices $\mathbf{p}_1, \mathbf{p}_2, \ldots, \mathbf{p}_n$. To achieve this we consider an arbitrary path $\mathbf{p}(t) = \mathbf{p} + t\mathbf{p}'$ in the space of the configurations $\mathbf{p} := (\mathbf{p}_1, \mathbf{p}_2, \ldots, \mathbf{p}_n)$, where $\mathbf{p}' := (\mathbf{p}'_1, \mathbf{p}'_2, \ldots, \mathbf{p}'_n)$. Based on Definition 7.4.1, Definition 7.4.2, and Lemma 7.4.3 we introduce $V(\mathbf{P}(t))$ as a function of t (with t being an arbitrary real with sufficiently small absolute value) via

$$\frac{1}{d!} \sum_{F_0 \subset F_1 \subset \cdots \subset F_{d-1}} \mathrm{sign}\left(\mathrm{conv}\{\mathbf{c}_{F_0}(t), \ldots, \mathbf{c}_{F_{d-1}}(t)\}\right) \det[\mathbf{c}_{F_0}(t), \ldots, \mathbf{c}_{F_{d-1}}(t)]$$

$$= \frac{1}{d!} \sum_{F_0 \subset F_1 \subset \cdots \subset F_{d-1}} \mathrm{sign}\left(\mathrm{conv}\{\mathbf{c}_{F_0}(t), \ldots, \mathbf{c}_{F_{d-1}}(t)\}\right) \mathbf{c}_{F_0}(t) \wedge \cdots \wedge \mathbf{c}_{F_{d-1}}(t),$$

where $\mathbf{c}_F(t) := \frac{1}{k}\sum_{j=1}^{k} \mathbf{p}_{i_j}(t)$ for any face $F = \mathrm{conv}\{\mathbf{p}_{i_1}, \ldots, \mathbf{p}_{i_k}\}$ of $\mathbf{P}$. Clearly, $V(\mathbf{P}(0)) = V(\mathbf{P})$. Moreover, evaluating the derivative $\frac{d}{dt}V(\mathbf{P}(t))$ of $V(\mathbf{P}(t))$ at $t = 0$, collecting terms, and using the anticommutativity of the wedge product we get that

$$\frac{d}{dt}V(\mathbf{P}(t))\,|_{t=0} = \frac{1}{d!}\sum_{i=1}^{n} \mathbf{N}_i \wedge \mathbf{p}'_i, \tag{7.12}$$

where each $\mathbf{N}_i$ is some linear combination of wedge products of $d-1$ vectors $\mathbf{p}_j$ with $\mathbf{p}_j$ and $\mathbf{p}_i$ sharing a common face.

Definition 7.4.4 *We call $\mathbf{N} := (\mathbf{N}_1, \mathbf{N}_2, \ldots, \mathbf{N}_n)$ the volume force of the d-dimensional convex polytope $\mathbf{P} \subset \mathbb{E}^d$ with n vertices.*

The following are some simple properties of the volume force. We leave the rather straightforward proofs to the reader.

Lemma 7.4.5 *Let $\mathbf{N} := (\mathbf{N}_1, \mathbf{N}_2, \ldots, \mathbf{N}_n)$ be the volume force of the d-dimensional convex polytope $\mathbf{P} \subset \mathbb{E}^d, d \geq 2$ with vertices $\mathbf{p}_1, \mathbf{p}_2, \ldots, \mathbf{p}_n$. Then the following hold.*
(1) Each $\mathbf{N}_i$ is only a function of the vertices that share a face with $\mathbf{p}_i$, but not $\mathbf{p}_i$ itself.
(2) Assume that the origin $\mathbf{o}$ of $\mathbb{E}^d$ is the barycenter of $\mathbf{P}$; moreover, let $T : \mathbb{E}^d \to \mathbb{E}^d$ be an orthogonal linear map satisfying $T(\mathbf{P}) = \mathbf{P}$. If $T(\mathbf{p}_i) = \mathbf{p}_j$, then $T(\mathbf{N}_i) = \mathbf{N}_j$.

For more details and examples on volume forces we refer the interested reader to the proper sections in [34].

7.4.3 Critical volume condition

Let $\mathbf{P} := \operatorname{conv}\{\mathbf{p}_1, \mathbf{p}_2, \ldots, \mathbf{p}_n\}$ be a d-dimensional convex polytope in $\mathbb{E}^d, d \geq 2$ with vertices $\mathbf{p} := (\mathbf{p}_1, \mathbf{p}_2, \ldots, \mathbf{p}_n)$. Let G be a graph defined on this vertex set $\mathbf{p}$. Here, G may or may not consist of the edges of $\mathbf{P}$. We think of the edges of G as defining those pairs of vertices of $\mathbf{P}$ that are constrained not to get closer. In the terminology of the geometry of rigid tensegrity frameworks each edge of G is a *strut*. (For more information on rigid tensegrity frameworks and the basic terminology used there we refer the interested reader to [222].)

Let $\mathbf{p}' := (\mathbf{p}_1', \mathbf{p}_2', \ldots, \mathbf{p}_n')$ be an *infinitesimal flex* of $G(\mathbf{p})$, where $G(\mathbf{p})$ refers to the realization of G over the point configuration $\mathbf{p}$. That is, for each edge (strut) $\{i, j\}$ of G we have

$$(\mathbf{p}_i - \mathbf{p}_j) \cdot (\mathbf{p}_i' - \mathbf{p}_j') \geq 0, \tag{7.13}$$

where "$\cdot$" denotes the standard inner product (also called the "dot product") in $\mathbb{E}^d$.

Let e denote the number of edges of G. Then the *rigidity matrix* $R(\mathbf{p})$ of $G(\mathbf{p})$ is the $e \times nd$ matrix whose row corresponding to the edge $\{i, j\}$ of G consists of the coordinates of d-dimensional vectors within a sequence of n vectors such that all the coordinates are zero except maybe the ones that correspond to the coordinates of the vectors $\mathbf{p}_i - \mathbf{p}_j$ and $\mathbf{p}_j - \mathbf{p}_i$ listed on the ith and jth position. Another way to introduce $R(\mathbf{p})$ is the following. Let $f : \mathbb{E}^{nd} \to \mathbb{E}^e$ be the map defined by $\mathbf{x} = (\mathbf{x}_1, \mathbf{x}_2, \ldots, \mathbf{x}_n) \to (\ldots, \|\mathbf{x}_i - \mathbf{x}_j\|^2, \ldots)$. Then it is immediate that $\frac{1}{2}\frac{d}{d\mathbf{x}}f|_{\mathbf{x}=\mathbf{p}} = R(\mathbf{p})$. Now, we can rewrite the inequalities of (7.13) in terms of the rigidity matrix $R(\mathbf{p})$ of $G(\mathbf{p})$ (using the usual matrix multiplication applied to $R(\mathbf{p})$ and the indicated column vector) as follows,

$$R(\mathbf{p})(\mathbf{p}')^T \geq 0, \tag{7.14}$$

where the inequality is meant for each coordinate.

For each edge $\{i, j\}$ of G, let ω_{ij} be a scalar. We collect all such scalars into a single row vector called the *stress* $\omega := (\ldots, \omega_{ij}, \ldots)$ corresponding to the rows of the matrix $R(\mathbf{p})$. Append the volume force $\mathbf{N} := (\mathbf{N}_1, \mathbf{N}_2, \ldots, \mathbf{N}_n)$ as the last row onto $R(\mathbf{p})$ to get a new matrix $\widehat{R}(\mathbf{p})$, which we call the *augmented rigidity matrix*. So, when performing the matrix multiplication $\widehat{R}(\mathbf{p})(\mathbf{p}')^T$, we find that the result is a column vector of length $e+1$ having $(\mathbf{p}_i - \mathbf{p}_j) \cdot (\mathbf{p}_i' - \mathbf{p}_j')$ on the position corresponding to the edge $\{i, j\}$ of G, and having $\sum_{k=1}^{n} \mathbf{N}_k \cdot \mathbf{p}_k'$ on the $(e + 1)$st position. Also, it is easy to see that

$$(\omega, 1)\widehat{R}(\mathbf{p}) = \left(\ldots, \sum_j \omega_{ij}(\mathbf{p}_i - \mathbf{p}_j) + \mathbf{N}_i, \ldots \right), \tag{7.15}$$

where each sum is taken over all $\mathbf{p}_j$ adjacent to $\mathbf{p}_i$ in G, and we collect d coordinates at a time.

Definition 7.4.6 *Let* $\mathbf{N} = (\mathbf{N}_1, \mathbf{N}_2, \dots, \mathbf{N}_n)$ *be the volume force of the d-dimensional convex polytope* $\mathbf{P} \subset \mathbb{E}^d, d \geq 2$ *with vertices* $\mathbf{p} = (\mathbf{p}_1, \mathbf{p}_2, \dots, \mathbf{p}_n)$. *We say that the stress* $\omega = (\dots, \omega_{ij}, \dots)$ *resolves* $\mathbf{N}$ *if for each i we have that* $\sum_j \omega_{ij}(\mathbf{p}_i - \mathbf{p}_j) + \mathbf{N}_i = \mathbf{o}$ *or, equivalently,* $(\omega, 1)\widehat{R}(\mathbf{p}) = \mathbf{o}$, *where* $\mathbf{o}$ *denotes the zero vector.*

Definition 7.4.7 *The d-dimensional convex polytope* $\mathbf{P} \subset \mathbb{E}^d, d \geq 2$ *and the graph G defined on the vertices of* $\mathbf{P}$ *satisfy the critical volume condition if the volume force* $\mathbf{N}$ *can be resolved by a stress* $\omega = (\dots, \omega_{ij}, \dots)$ *such that for each edge $\{i, j\}$ of G, $\omega_{ij} < 0$.*

Theorem 7.4.8 *Let the d-dimensional convex polytope* $\mathbf{P} \subset \mathbb{E}^d, d \geq 2$ *and the strut graph G, defined on the vertices of* $\mathbf{P}$, *satisfy the critical volume condition. Moreover, let* $\mathbf{p}' = (\mathbf{p}'_1, \mathbf{p}'_2, \dots, \mathbf{p}'_n)$ *be an infinitesimal flex of the strut framework $G(\mathbf{p})$ (i.e., let $\mathbf{p}'$ satisfy (7.13)). Then*

$$\frac{d}{dt} V\left(\mathbf{P}(t)\right)|_{t=0} = \frac{1}{d!} \sum_{i=1}^{n} \mathbf{N}_i \wedge \mathbf{p}'_i \geq 0$$

with equality if and only if $(\mathbf{p}_i - \mathbf{p}_j) \cdot (\mathbf{p}'_i - \mathbf{p}'_j) = 0$ *for each edge $\{i, j\}$ of G.*

Proof: The assumptions, (7.15), the associativity of matrix multiplication, and (7.12) imply in a straightforward way that

$$0 = \mathbf{o} \cdot \mathbf{p}' = (\omega, 1)\widehat{R}(\mathbf{p})(\mathbf{p}')^T = \sum_{\{i,j\}} \omega_{ij}(\mathbf{p}_i - \mathbf{p}_j) \cdot (\mathbf{p}'_i - \mathbf{p}'_j) + \sum_{i=1}^{n} \mathbf{N}_i \cdot \mathbf{p}'_i$$

$$= \sum_{\{i,j\}} \omega_{ij}(\mathbf{p}_i - \mathbf{p}_j) \cdot (\mathbf{p}'_i - \mathbf{p}'_j) + \sum_{i=1}^{n} \mathbf{N}_i \wedge \mathbf{p}'_i \leq \sum_{i=1}^{n} \mathbf{N}_i \wedge \mathbf{p}'_i = \frac{d}{dt} V\left(\mathbf{P}(t)\right)|_{t=0},$$

where $\mathbf{N}_i$ is regarded as a d-dimensional vector so that $\mathbf{N}_i \wedge \mathbf{p}'_i$ can be interpreted as the standard inner product $\mathbf{N}_i \cdot \mathbf{p}_i$, with appropriate identification of bases. We clearly get equality if and only if $(\mathbf{p}_i - \mathbf{p}_j) \cdot (\mathbf{p}'_i - \mathbf{p}'_j) = 0$ for each edge $\{i, j\}$ of G. $\qquad\square$

7.4.4 Strictly locally volume expanding convex polytopes

The following definition recalls standard terminology from the theory of rigid tensegrity frameworks. (See [105] for more information.) Consider now just the *bar graph* $\overline{G}$, which is the graph G with all the struts changed to *bars*, and take its realization $\overline{G}(\mathbf{p})$ sitting over the point configuration $\mathbf{p} = (\mathbf{p}_1, \mathbf{p}_2, \dots, \mathbf{p}_n)$. (Here bars mean edges whose lengths are constrained not to change.) We say that the infinitesimal motion $\mathbf{p}' = (\mathbf{p}'_1, \mathbf{p}'_2, \dots, \mathbf{p}'_n)$ is an *infinitesimal flex* of $\overline{G}(\mathbf{p})$ if for each edge (bar) $\{i, j\}$ of $\overline{G}$, we have

$$(\mathbf{p}_i - \mathbf{p}_j) \cdot (\mathbf{p}'_i - \mathbf{p}'_j) = 0.$$

This is the same as saying $R(\mathbf{p})(\mathbf{p}')^T = \mathbf{o}$ for the rigidity matrix $R(\mathbf{p})$.

Definition 7.4.9 *We say that* $\mathbf{p}'$ *is a trivial infinitesimal flex if* $\mathbf{p}'$ *is a (directional) derivative of an isometric motion of* $\mathbb{E}^d, d \geq 2$. *We say that* $G(\mathbf{p})$ *(resp.,* $\overline{G}(\mathbf{p})$*) is infinitesimally rigid if* $G(\mathbf{p})$ *(resp.,* $\overline{G}(\mathbf{p})$*) has only trivial infinitesimal flexes.*

Theorem 7.4.10 *Let the d-dimensional convex polytope* $\mathbf{P} \subset \mathbb{E}^d, d \geq 2$ *and the strut graph* G, *defined on the vertices of* $\mathbf{P}$, *satisfy the critical volume condition and assume that the bar framework* $\overline{G}(\mathbf{p})$ *is infinitesimally rigid. Then*

$$\frac{d}{dt} V\left(\mathbf{P}(t)\right)|_{t=0} = \frac{1}{d!} \sum_{i=1}^{n} \mathbf{N}_i \wedge \mathbf{p}'_i > 0$$

for every non-trivial infinitesimal flex $\mathbf{p}' = (\mathbf{p}'_1, \mathbf{p}'_2, \ldots, \mathbf{p}'_n)$ *of the strut framework* $G(\mathbf{p})$.

Proof: By Theorem 7.4.8 we have that $\frac{d}{dt} V\left(\mathbf{P}(t)\right)|_{t=0} = \frac{1}{d!} \sum_{i=1}^{n} \mathbf{N}_i \wedge \mathbf{p}'_i \geq 0$. If $\frac{d}{dt} V\left(\mathbf{P}(t)\right)|_{t=0} = 0$, then applying Theorem 7.4.8 again, $\mathbf{p}' = (\mathbf{p}'_1, \mathbf{p}'_2, \ldots, \mathbf{p}'_n)$ must be an infinitesimal flex of the bar framework $\overline{G}(\mathbf{p})$. However, then by the infinitesimal rigidity of $\overline{G}(\mathbf{p})$, this would imply that $\mathbf{p}'$ is trivial. Thus, $\frac{d}{dt} V\left(\mathbf{P}(t)\right)|_{t=0} > 0$. $\square$

The following definition leads us to the core part of this section.

Definition 7.4.11 *Let* $\mathbf{P} \subset \mathbb{E}^d, d \geq 2$ *be a d-dimensional convex polytope and let* G *be a strut graph defined on the vertices* $\mathbf{p} = (\mathbf{p}_1, \mathbf{p}_2, \ldots, \mathbf{p}_n)$ *of* $\mathbf{P}$. *We say that* $\mathbf{P}$ *is strictly locally volume expanding over* G, *if there is an* $\epsilon > 0$ *with the following property. For every* $\mathbf{q} = (\mathbf{q}_1, \mathbf{q}_2, \ldots, \mathbf{q}_n)$ *satisfying*

$$\|\mathbf{p}_i - \mathbf{q}_i\| < \epsilon \text{ for all } i = 1, \ldots, n \tag{7.16}$$

and

$$\|\mathbf{p}_i - \mathbf{p}_j\| \leq \|\mathbf{q}_i - \mathbf{q}_j\| \text{ for each edge } \{i, j\} \text{ of } G, \tag{7.17}$$

we have $V(\mathbf{P}) \leq V(\mathbf{Q})$ *(where* $V(\mathbf{Q})$ *is defined via (7.10) and (7.11) substituting* $\mathbf{q}$ *for* $\mathbf{p}$*) with equality only when* $\mathbf{P}$ *is congruent to* $\mathbf{Q}$, *where* $\mathbf{Q}$ *is the polytope generated by the simplices of the barycenters in (7.10) using* $\mathbf{q}$ *instead of* $\mathbf{p}$.

Theorem 7.4.12 *Let the d-dimensional convex polytope* $\mathbf{P} \subset \mathbb{E}^d, d \geq 2$ *and the strut graph* G, *defined on the vertices of* $\mathbf{P}$, *satisfy the critical volume condition and assume that the bar framework* $\overline{G}(\mathbf{p})$ *is infinitesimally rigid. Then* $\mathbf{P}$ *is strictly locally volume expanding over* G.

Proof: The inequalities (7.17) define a semialgebraic set X in the space of all configurations $\{(\mathbf{q}_1, \mathbf{q}_2, \ldots, \mathbf{q}_n) | \mathbf{q}_i \in \mathbb{E}^d, i = 1, \ldots, n\}$. Suppose there is no ϵ as in the conclusion. Add $V(\mathbf{P}) \geq V(\mathbf{Q})$ to the constraints defining X. By Wallace [245] (see [105]) there is an *analytic* path $\mathbf{p}(t) = (\mathbf{p}_1(t), \mathbf{p}_2(t), \ldots, \mathbf{p}_n(t)), 0 \leq t < 1$, with $\mathbf{p}(0) = \mathbf{p}$ and $\mathbf{p}(t) \in X$, $\mathbf{p}(t)$ not congruent to $\mathbf{p}(0)$ for $0 < t < 1$. So,

$$\|\mathbf{p}_i - \mathbf{p}_j\| \leq \|\mathbf{p}_i(t) - \mathbf{p}_j(t)\| \text{ for each edge } \{i, j\} \text{ of } G \text{ and} \tag{7.18}$$

$$V(\mathbf{P}) \geq V(\mathbf{P}(t)) \text{ for } 0 \leq t < 1. \tag{7.19}$$

Then after suitably adjusting $\mathbf{p}(t)$ by congruences (as in [105] as well as [107]) we can define

$$\mathbf{p}' := \frac{d^k \mathbf{p}(t)}{dt^k}|_{t=0}$$

for the smallest k that makes $\mathbf{p}'$ a non-trivial infinitesimal flex. (Such k exists by the argument in [105] as well as [107]).

Because (7.18) holds we see that $\mathbf{p}'$ is a non-trivial infinitesimal flex of $G(\mathbf{p})$ and (7.19) implies that

$$\frac{d}{dt} V(\mathbf{P}(t))|_{t=0} \leq 0.$$

But this contradicts Theorem 7.4.10, finishing the proof of Theorem 7.4.12.
□

7.4.5 From critical volume condition and infinitesimal rigidity to uniform stability of sphere packings

Here we start with the assumptions of Theorem 1.7.3 and apply Theorem 7.4.12 to each $\mathbf{P}_i$ and $G_{\mathcal{P}}$ restricted to the vertices of $\mathbf{P}_i$, $1 \leq i \leq m$. Then let $\epsilon_0 > 0$ be the smallest $\epsilon > 0$ guaranteed by the strict locally volume expanding property of Theorem 7.4.12. All but a finite number of tiles are fixed. The tiles that are free to move are confined to a region of fixed volume in $\mathbb{E}^d, d \geq 2$. Each $\mathbf{P}_i$ is strictly locally volume expanding, therefore the volume of each of the tiles must be fixed. But the strict condition implies that the motion of each tile must be an isometry. Because the tiling is face-to-face and the vertices are given by $G_{\mathcal{P}}$ we conclude inductively (on the number of tiles) that each vertex of $G_{\mathcal{P}}$ must be fixed. Thus, $\mathcal{P}$ is uniformly stable with respect to ϵ_0 introduced above, finishing the proof of Theorem 1.7.3.

8

Selected Proofs on Finite Packings of Translates of Convex Bodies

8.1 Proof of Theorem 2.2.1

8.1.1 Monotonicity of a special integral function

Lemma 8.1.1 *Let $f : [0,1] \to \mathbb{R}$ be a function such that f is positive and monotone increasing on $(0,1]$; moreover, $f(x) = (g(x))^k$ for some concave function $g : [0,1] \to \mathbb{R}$, where k is a positive integer. Then*

$$F(y) := \frac{1}{f(y)} \int_0^y f(x)dx$$

is strictly monotone increasing on $(0,1]$.

Proof: Without loss of generality we may assume that f is differentiable. So, to prove that $F(y) := \frac{1}{f(y)} \int_0^y f(x)dx$ is strictly monotone increasing, it is sufficient to show that $\frac{d}{dy}F > 0$ or equivalently that $\int_0^y f(x)dx < \frac{(f(y))^2}{f'(y)}$. From now on, let $0 < y < 1$ be fixed (with $f'(y) > 0$).

As $f = g^k$ for some concave g therefore the linear function $l(x) = b_1 + b_2(x - y)$ with $b_1 = (f(y))^{\frac{1}{k}}$ and $b_2 = \frac{f'(y)}{k(f(y))^{\frac{k-1}{k}}}$ satisfies the inequality $g(x) \leq l(x)$ for all $0 \leq x \leq 1$, and so we have that $f(x) \leq (l(x))^k$ holds for all $0 \leq x \leq 1$. Thus, for all $0 \leq x \leq 1$ we have

$$f(x) \leq \left((f(y))^{\frac{1}{k}} + \frac{f'(y)}{k(f(y))^{\frac{k-1}{k}}}(x - y) \right)^k = f(y) \left(1 + \frac{f'(y)}{kf(y)}(x - y) \right)^k .$$

$$(8.1)$$

By integration we get

$$\int_0^y f(x)dx \leq \int_0^y f(y) \left(1 + \frac{f'(y)}{kf(y)}(x - y) \right)^k dx$$

$$= \frac{k}{k+1} \frac{(f(y))^2}{f'(y)} \left(1 - \left(1 - \frac{y f'(y)}{k f(y)} \right)^{k+1} \right). \tag{8.2}$$

Now, because the first factor of (8.2) is strictly between 0 and 1, it is sufficient to show that the last factor is at most 1; that is, we are left to show the inequality

$$0 \le \left(1 - \frac{y f'(y)}{k f(y)} \right)^{k+1} \tag{8.3}$$

Suppose that (8.3) is not true; then $\left(1 - \frac{y f'(y)}{k f(y)} \right)^{k+1} < 0$. Let $G(x) :=$ $\left(1 + \frac{f'(y)}{k f(y)} (x - y) \right)^{k+1}$. As $G(y) = 1$ and by assumption $G(0) < 0$, therefore there must be an $0 < x_0 < y$ such that $G(x_0) = \left(1 + \frac{f'(y)}{k f(y)} (x_0 - y) \right)^{k+1} = 0$. But then this and (8.1) imply in a straightforward way that $f(x_0) \le f(y) \left(1 + \frac{f'(y)}{k f(y)} (x_0 - y) \right)^{k} = 0$. However, by the assumptions of Lemma 8.1.1 we have that $f(x_0) > 0$, a contradiction. This completes our proof of Lemma 8.1.1. □

8.1.2 A proof by slicing via the Brunn–Minkowski inequality

Let the convex body $\mathbf{K}$ be positioned in $\mathbb{E}^d$ such that the hyperplane $\{\mathbf{x} \in \mathbb{E}^d \mid \langle \mathbf{x}, \mathbf{v} \rangle = -1\}$ with normal vector $\mathbf{v}$ is a supporting hyperplane for $\mathbf{K}$ and the non-overlapping translates $\mathbf{t}_1 + \mathbf{K}, \ldots, \mathbf{t}_k + \mathbf{K}$ are all touching $\mathbf{K}$ and (together with $\mathbf{K}$) are all lying in the closed halfspace $\{\mathbf{x} \in \mathbb{E}^d \mid \langle \mathbf{x}, \mathbf{v} \rangle \ge -1\}$. Now, due to the well-known fact that by replacing $\mathbf{K}$ with $\frac{1}{2}(\mathbf{K} + -(\mathbf{K}))$ and performing the same symmetrization for each of the translates $\mathbf{t}_1 + \mathbf{K}, \ldots, \mathbf{t}_k + \mathbf{K}$ one preserves the packing property, touching pairs, and one-sidedness, we may assume that $\mathbf{K}$ is in fact, a centrally symmetric convex body of $\mathbb{E}^d$ say, it is $\mathbf{o}$-symmetric, where $\mathbf{o}$ stands for the origin of $\mathbb{E}^d$. Moreover, as in the classical proof for the Hadwiger number [151], we use that $\bigcup_{i=0}^k (\mathbf{t}_i + \mathbf{K}) \subset 3\mathbf{K}$, where $\mathbf{t}_0 = \mathbf{o}$. Furthermore, let the family $\mathbf{t}_0 + \mathbf{K}, \mathbf{t}_1 + \mathbf{K}, \ldots, \mathbf{t}_k + \mathbf{K}$ be scaled so that the normal vector $\mathbf{v}$ is a unit vector (i.e., $\|\mathbf{v}\| = 1$). Next, let $H_x := \{\mathbf{p} \in \mathbb{E}^d \mid \langle \mathbf{p}, \mathbf{v} \rangle = x\}$ for $x \in \mathbb{R}$. Then clearly, $\mathbf{K}$ is between the hyperplanes H_{-1} and H_1 touching both, and the translates $\mathbf{t}_1 + \mathbf{K}, \ldots, \mathbf{t}_k + \mathbf{K}$ (together with $\mathbf{K} = \mathbf{t}_0 + \mathbf{K}$) all lie between the hyperplanes H_{-1} and H_3. Obviously, $\int_{-1}^1 \text{vol}_{d-1}(\mathbf{K} \cap H_x) \, dx = \text{vol}_d(\mathbf{K})$, where $\text{vol}_d(\cdot)$ (resp., $\text{vol}_{d-1}(\cdot)$) denotes the d-dimensional (resp., $d-1$-dimensional) volume measure. Also, it follows from the given setup in a straightforward way that

$$\int_{-1}^3 \text{vol}_{d-1} \left(H_x \cap \bigcup_{i=0}^k (\mathbf{t}_i + \mathbf{K}) \right) dx = (k+1)\text{vol}_d(\mathbf{K}). \tag{8.4}$$

Our goal is to write the integral in (8.4) as a sum of two integrals from -1 to 0 and from 0 to 3, and estimate them separately.

First, notice that

$$\int_0^3 \text{vol}_{d-1}\left(H_x \cap \bigcup_{i=0}^k (t_i + \mathbf{K})\right) dx \leq \int_0^3 \text{vol}_{d-1}\left(H_x \cap (3\mathbf{K})\right) dx = \frac{3^d}{2}\text{vol}_d(\mathbf{K}).$$
$$(8.5)$$

Second, notice that

$$\int_{-1}^0 \text{vol}_{d-1}\left(H_x \cap \bigcup_{i=0}^k (t_i + \mathbf{K})\right) dx = \sum_{i=0}^k \int_{-1}^0 \text{vol}_{d-1}\left(H_x \cap (t_i + \mathbf{K})\right) dx$$
$$(8.6)$$

$$= \sum_{i=0}^k \int_0^1 \text{vol}_{d-1}\left(\mathbf{K} \cap (-t_i + H_{x-1})\right) dx = \sum_{0 \leq a_i \leq 1} \int_0^{1-a_i} f(x) dx, \quad (8.7)$$

where $f(x) := \text{vol}_{d-1}\left(\mathbf{K} \cap H_{x-1}\right), 0 \leq x \leq 1$ and $a_i := \langle \mathbf{v}, t_i \rangle, 0 \leq i \leq k$. We note that $a_i \geq 0$ for all $0 \leq i \leq k$ (and for some j we have that $a_j \geq 1$). Moreover, f is positive and monotone increasing on $(0, 1]$, and by the Brunn–Minkowski inequality (see, e.g., [85]) the function $f^{\frac{1}{d-1}}$ is concave (for all $d \geq 2$). Thus, Lemma 8.1.1 implies that

$$\sum_{0 \leq a_i \leq 1} \int_0^{1-a_i} f(x) dx \leq \sum_{0 \leq a_i \leq 1} \left(\frac{f(1-a_i)}{f(1)} \int_0^1 f(x) dx\right) \quad (8.8)$$

$$= \frac{\int_0^1 f(x) dx}{f(1)} \sum_{0 \leq a_i \leq 1} f(1-a_i) = \frac{\int_0^1 f(x) dx}{f(1)} \sum_{0 \leq a_i \leq 1} \text{vol}_{d-1}\left(\mathbf{K} \cap H_{-a_i}\right) \quad (8.9)$$

$$= \frac{\int_0^1 f(x) dx}{f(1)} \sum_{i=0}^k \text{vol}_{d-1}\left((t_i + \mathbf{K}) \cap H_0\right) \quad (8.10)$$

$$= \frac{\int_0^1 f(x) dx}{f(1)} \text{vol}_{d-1}\left(H_0 \cap \bigcup_{i=0}^k (t_i + \mathbf{K})\right) \leq \frac{\int_0^1 f(x) dx}{f(1)} \text{vol}_{d-1}\left(H_0 \cap (3\mathbf{K})\right)$$
$$(8.11)$$

$$= \frac{1}{2}\text{vol}_d(\mathbf{K}) \frac{1}{\text{vol}_{d-1}(H_0 \cap \mathbf{K})} \text{vol}_{d-1}\left(H_0 \cap (3\mathbf{K})\right) = \frac{3^{d-1}}{2}\text{vol}_d(\mathbf{K}). \quad (8.12)$$

Hence, (8.4), (8.5), (8.6), (8.7), (8.8), (8.9), (8.10), (8.11), and (8.12) yield that

$$(k+1)\text{vol}_d(\mathbf{K}) \leq \frac{3^d}{2}\text{vol}_d(\mathbf{K}) + \frac{3^{d-1}}{2}\text{vol}_d(\mathbf{K}),$$

and so, $k \leq 2 \cdot 3^{d-1} - 1$ as claimed in Theorem 2.2.1.

To prove that equality can only be reached for d-dimensional affine cubes, notice first that the equality in (8.8) and the strict monotonicity of Lemma 8.1.1 imply that for all a_i with $0 \le a_i < 1$ we have $a_i = 0$ and for all $a_i = 1$ we have $f(1 - a_i) = 0$. Taking into account the equality in (8.11), we get that translates of $H_0 \cap \mathbf{K}$ must tile $H_0 \cap (3\mathbf{K})$. Hence, [151] yields that $H_0 \cap \mathbf{K}$ as well as $H_0 \cap (3\mathbf{K})$ are $(d-1)$-dimensional affine cubes. Also, there is only the obvious way to tile $H_0 \cap (3\mathbf{K})$ by 3^{d-1} translates of $H_0 \cap \mathbf{K}$, so the set of the translation vectors $\{\mathbf{t}_i \mid \mathbf{t}_i \in H_0\}$ is $\mathbf{o}$-symmetric. But then the $\left((3^{d-1} - 1) + 1\right) + 2\left((2 \cdot 3^{d-1} - 1) - (3^{d-1} - 1)\right) = 3^d$ translates

$$\{\mathbf{t}_i + \mathbf{K} \mid \mathbf{t}_i \in H_0\} \cup \{\mathbf{t}_i + \mathbf{K} \mid \mathbf{t}_i \notin H_0\} \cup \{-\mathbf{t}_i + \mathbf{K} \mid \mathbf{t}_i \notin H_0\}$$

of $\mathbf{K}$ form a packing in $3\mathbf{K}$. Thus, the Hadwiger number of $\mathbf{K}$ is $3^d - 1$ and so, using [151] we get that $\mathbf{K}$ is indeed a d-dimensional affine cube. This completes the proof of Theorem 2.2.1.

8.2 Proof of Theorem 2.4.3

Let $\mathbf{C}_n := \{\mathbf{c}_1, \mathbf{c}_2, \dots, \mathbf{c}_n\}$ and assume that the inequality stated in Theorem 2.4.3 does not hold. Then there is an $\epsilon > 0$ such that

$$\text{vol}_d(\mathbf{C}_n + 2\mathbf{K_o}) = \frac{n\,\text{vol}_d(\mathbf{K_o})}{\delta(\mathbf{K_o})} - \epsilon. \tag{8.13}$$

Let $\Lambda \subset \mathbb{E}^d$ be a d-dimensional packing lattice of $\mathbf{C}_n + 2\mathbf{K_o}$ such that $\mathbf{C}_n + 2\mathbf{K_o}$ is contained in the fundamental parallelotope $\mathbf{P}$ of Λ. For each $\lambda > 0$ let $\mathbf{Q}_\lambda$ denote the d-dimensional cube of edge length 2λ centered at the origin $\mathbf{o}$ of $\mathbb{E}^d$ having edges parallel to the corresponding coordinate axes of $\mathbb{E}^d$. Obviously, there is a constant $\mu > 0$ depending on $\mathbf{P}$ only such that for each $\lambda > 0$ there is a subset $L_\lambda \subset \Lambda$ with $\mathbf{Q}_\lambda \subset L_\lambda + \mathbf{P}$ and $L_\lambda + 2\mathbf{P} \subset \mathbf{Q}_{\lambda+\mu}$. Moreover, let $\mathcal{P}_n(\mathbf{K_o})$ be the family of all possible packings of $n > 1$ translates of the $\mathbf{o}$-symmetric convex body $\mathbf{K_o}$ in $\mathbb{E}^d$. The definition of $\delta(\mathbf{K_o})$ implies that for each $\lambda > 0$ there exists a packing in the family $\mathcal{P}_{m(\lambda)}(\mathbf{K_o})$ with centers at the points of $\mathbf{C}_{m(\lambda)}$ such that $\mathbf{C}_{m(\lambda)} + \mathbf{K_o} \subset \mathbf{Q}_\lambda$ and

$$\lim_{\lambda \to \infty} \frac{m(\lambda)\text{vol}_d(\mathbf{K_o})}{\text{vol}_d(\mathbf{Q}_\lambda)} = \delta(\mathbf{K_o}).$$

As $\lim_{\lambda \to \infty} \frac{\text{vol}_d(\mathbf{Q}_{\lambda+\mu})}{\text{vol}_d(\mathbf{Q}_\lambda)} = 1$, therefore there exist $\xi > 0$ and a packing in the family $\mathcal{P}_{m(\xi)}(\mathbf{K_o})$ with centers at the points of $\mathbf{C}_{m(\xi)}$ and with $\mathbf{C}_{m(\xi)} + \mathbf{K_o} \subset \mathbf{Q}_\xi$ such that

$$\frac{\text{vol}_d(\mathbf{P})\delta(\mathbf{K_o})}{\text{vol}_d(\mathbf{P}) + \epsilon} < \frac{m(\xi)\text{vol}_d(\mathbf{K_o})}{\text{vol}_d(\mathbf{Q}_{\xi+\mu})} \tag{8.14}$$

and

$$\frac{n\mathrm{vol}_d(\mathbf{K_o})}{\mathrm{vol}_d(\mathbf{P}) + \epsilon} < \frac{n\mathrm{vol}_d(\mathbf{K_o})\mathrm{card}(L_\xi)}{\mathrm{vol}_d(\mathbf{Q}_{\xi+\mu})}. \tag{8.15}$$

Now, for each $\mathbf{x} \in \mathbf{P}$ we define a packing of $n(\mathbf{x})$ translates of the o-symmetric convex body $\mathbf{K_o}$ in $\mathbb{E}^d$ with centers at the points of

$$\mathbf{C}_{n(\mathbf{x})} := \{\mathbf{x} + L_\xi + \mathbf{C}_n\} \cup \{\mathbf{y} \in \mathbf{C}_{m(\xi)} \mid \mathbf{y} \notin \mathbf{x} + L_\xi + \mathbf{C}_n + \mathrm{int}(2\mathbf{K_o})\}.$$

Clearly, $\mathbf{C}_{n(\mathbf{x})} + \mathbf{K_o} \subset \mathbf{Q}_{\xi+\mu}$. As a next step we introduce the (characteristic) function $\chi_{\mathbf{y}} : \mathbf{P} \to \mathbb{R}$ as follows: $\chi_{\mathbf{y}}(\mathbf{x}) := 1$ if $\mathbf{y} \notin \mathbf{x} + L_\xi + \mathbf{C}_n + \mathrm{int}(2\mathbf{K_o})$ and $\chi_{\mathbf{y}}(\mathbf{x}) := 0$ for any other $\mathbf{x} \in \mathbf{P}$. Thus,

$$\int_{\mathbf{x} \in \mathbf{P}} n(\mathbf{x}) \, d\mathbf{x} = \int_{\mathbf{x} \in \mathbf{P}} \left(n \, \mathrm{card}(L_\xi) + \sum_{\mathbf{y} \in \mathbf{C}_{m(\xi)}} \chi_{\mathbf{y}}(\mathbf{x}) \right) d\mathbf{x}$$

$$= n\mathrm{vol}_d(\mathbf{P})\mathrm{card}(L_\xi) + m(\xi) \left(\mathrm{vol}_d(\mathbf{P}) - \mathrm{vol}_d(\mathbf{C}_n + 2\mathbf{K_o}) \right).$$

Hence, there is a point $\mathbf{p} \in \mathbf{P}$ with

$$n(\mathbf{p}) \geq m(\xi) \left(1 - \frac{\mathrm{vol}_d(\mathbf{C}_n + 2\mathbf{K_o})}{\mathrm{vol}_d(\mathbf{P})} \right) + n \, \mathrm{card}(L_\xi)$$

and so,

$$\frac{n(\mathbf{p})\mathrm{vol}_d(\mathbf{K_o})}{\mathrm{vol}_d(\mathbf{Q}_{\xi+\mu})}$$

$$\geq \frac{m(\xi)\mathrm{vol}_d(\mathbf{K_o})}{\mathrm{vol}_d(\mathbf{Q}_{\xi+\mu})} \left(1 - \frac{\mathrm{vol}_d(\mathbf{C}_n + 2\mathbf{K_o})}{\mathrm{vol}_d(\mathbf{P})} \right) + \frac{n\mathrm{vol}_d(\mathbf{K_o})\mathrm{card}(L_\xi)}{\mathrm{vol}_d(\mathbf{Q}_{\xi+\mu})}. \tag{8.16}$$

Thus, (8.16), (8.15), (8.14), and (8.13) imply in a straightforward way that

$$\frac{n(\mathbf{p})\mathrm{vol}_d(\mathbf{K_o})}{\mathrm{vol}_d(\mathbf{Q}_{\xi+\mu})}$$

$$> \frac{\mathrm{vol}_d(\mathbf{P})\delta(\mathbf{K_o})}{\mathrm{vol}_d(\mathbf{P}) + \epsilon} \left(1 - \frac{\mathrm{vol}_d(\mathbf{C}_n + 2\mathbf{K_o})}{\mathrm{vol}_d(\mathbf{P})} \right) + \frac{n\mathrm{vol}_d(\mathbf{K_o})}{\mathrm{vol}_d(\mathbf{P}) + \epsilon} = \delta(\mathbf{K_o}). \tag{8.17}$$

As $\mathbf{C}_{n(\mathbf{p})} + \mathbf{K_o} \subset \mathbf{Q}_{\xi+\mu}$, therefore (8.17) leads to the existence of a packing by translates of $\mathbf{K_o}$ in $\mathbb{E}^d$ with density strictly larger than $\delta(\mathbf{K_o})$, a contradiction. This finishes the proof of Theorem 2.4.3.

9

Selected Proofs on Illumination and Related Topics

9.1 Proof of Corollary 3.4.2 Using Rogers' Classical Theorem on Economical Coverings

The proof is based on the following classical theorem of Rogers [217]. We do not prove it here, but instead refer the interested reader to the numerous resources on that, in particular to [219], [1], [143], and [131]. Let $\vartheta(\mathbf{K})$ denote the infimum of the densities of coverings of $\mathbb{E}^d$ by translates of the convex body $\mathbf{K}$.

Theorem 9.1.1 *For any convex body $\mathbf{K} \subset \mathbb{E}^d, d \geq 2$ there exists a covering of $\mathbb{E}^d$ by translates of $\mathbf{K}$ with density not exceeding $d \ln d + d \ln \ln d + 5d$; that is, we have $\vartheta(\mathbf{K}) \leq d \ln d + d \ln \ln d + 5d$. Moreover, for sufficiently large d, $5d$ can be replaced by $4d$.*

Now, the proof of Corollary 3.4.2 goes as follows. Clearly, for each $\epsilon > 0$ we can choose a discrete set $T \subset \mathbb{E}^d$ such that

$$\bigcup_{t \in T} t + \mathrm{int}\mathbf{K} = \mathbb{E}^d \text{ and } \frac{\sum_{t \in T \cap \mathbf{C}} \mathrm{vol}_d(t + \mathrm{int}\mathbf{K})}{\mathrm{vol}_d(\mathbf{C})} \leq \vartheta(\mathrm{int}\mathbf{K}) + \epsilon$$

holds for every sufficiently large cube $\mathbf{C}$ (centered at the origin $\mathbf{o}$ of $\mathbb{E}^d$). Let $N(\mathbf{x}) := \mathrm{card}\,(T \cap \mathbf{x} + (\mathbf{K} - \mathrm{int}\mathbf{K}))$ with $\mathbf{x} \in \mathbb{E}^d$, where $\mathrm{card}(\cdot)$ refers to the cardinality of the corresponding set.

Lemma 9.1.2
$$I(\mathbf{K}) \leq N(\mathbf{x})$$

for all $\mathbf{x} \in \mathbb{E}^d$.

Proof: It is easy to see that $(t + \mathrm{int}\mathbf{K}) \cap (\mathbf{x} + \mathbf{K}) \neq \emptyset$ holds if and only if $t \in \mathbf{x} + (\mathbf{K} - \mathrm{int}\mathbf{K})$. Hence, as $\cup_{t \in T} t + \mathrm{int}\mathbf{K} = \mathbb{E}^d$, therefore

$$\mathbf{x} + \mathbf{K} \subset \bigcup_{t \in T \cap \mathbf{x} + (\mathbf{K} - \mathrm{int}\mathbf{K})} t + \mathrm{int}\mathbf{K},$$

K. Bezdek, *Classical Topics in Discrete Geometry*, CMS Books in Mathematics, DOI 10.1007/978-1-4419-0600-7_9, © Springer Science+Business Media, LLC 2010

finishing the proof of Lemma 9.1.2.   □

As by assumption we have that

$$\operatorname{card}(T \cap \mathbf{C}) \leq \frac{\operatorname{vol}_d(\mathbf{C})}{\operatorname{vol}_d(\operatorname{int}\mathbf{K})}(\vartheta(\operatorname{int}\mathbf{K}) + \epsilon),$$

therefore one can easily verify that

$$\frac{1}{\operatorname{vol}_d(\mathbf{C})} \int_{\mathbf{C}} N(\mathbf{x})d\mathbf{x} \leq \frac{\operatorname{vol}_d(\mathbf{K} - \operatorname{int}\mathbf{K})}{\operatorname{vol}_d(\operatorname{int}\mathbf{K})}(\vartheta(\operatorname{int}\mathbf{K}) + 2\epsilon)$$

holds for every sufficiently large cube $\mathbf{C}$. The latter inequality together with Lemma 9.1.2 implies the existence of $\mathbf{x}_0 \in \mathbb{E}^d$ satisfying

$$I(\mathbf{K}) \leq N(\mathbf{x}_0) \leq \frac{\operatorname{vol}_d(\mathbf{K} - \operatorname{int}\mathbf{K})}{\operatorname{vol}_d(\operatorname{int}\mathbf{K})}(\vartheta(\operatorname{int}\mathbf{K}) + 2\epsilon).$$

As $I(\mathbf{K})$ is an integer, by choosing ϵ sufficiently small, we can guarantee that

$$I(\mathbf{K}) \leq \frac{\operatorname{vol}_d(\mathbf{K} - \operatorname{int}\mathbf{K})}{\operatorname{vol}_d(\operatorname{int}\mathbf{K})}\vartheta(\operatorname{int}\mathbf{K}).$$

This together with Theorem 9.1.1 completes the proof of Corollary 3.4.2.

9.2 Proof of Theorem 3.5.2 via the Gauss Map

Recall the following statement published in [52] that gives a reformulation of the X-ray number of a convex body in terms of its Gauss map.

Lemma 9.2.1 *Let* $\mathbf{M}$ *be a convex body in* $\mathbb{E}^d, d \geq 3$, *and let* $\mathbf{b} \in \operatorname{bd}\mathbf{M}$ *be given; moreover, let* F *denote (any of) the face(s) of* $\mathbf{M}$ *of smallest dimension containing* $\mathbf{b}$. *Then* $\mathbf{b}$ *is X-rayed along the line* L *if and only if* $L^\perp \cap \nu(F) = \emptyset$, *where* $L^\perp$ *denotes the hyperplane orthogonal to* L *and passing through the origin* $\mathbf{o}$ *of* $\mathbb{E}^d$ *and* $\nu(F)$ *denotes the Gauss image of* F. *Moreover,* $X(\mathbf{M})$ *is the smallest number of* $(d-2)$-*dimensional great spheres of* $\mathbb{S}^{d-1}$ *with the property that the Gauss image of each face of* $\mathbf{M}$ *is disjoint from at least one of the given great spheres.*

Let $\{\mathbf{p}_1, -\mathbf{p}_1, \ldots, \mathbf{p}_m, -\mathbf{p}_m\}$ be the family of pairwise antipodal points in $\mathbb{S}^{d-1}$ with covering radius R. Moreover, let $B_i \subset \mathbb{S}^{d-1}$ be the union of the two $(d-1)$-dimensional closed spherical balls of radius R centered at the points $\mathbf{p}_i$ and $-\mathbf{p}_i$ in $\mathbb{S}^{d-1}$, $1 \leq i \leq m$. Finally, let S_i be the $(d-2)$-dimensional great sphere of $\mathbb{S}^{d-1}$ whose hyperplane is orthogonal to the diameter of $\mathbb{S}^{d-1}$ with endpoints $\mathbf{p}_i$ and $-\mathbf{p}_i$, $1 \leq i \leq m$. Based on Lemma 9.2.1 it is sufficient to show that the Gauss image of each face of $\mathbf{K}$ is disjoint from at least one of the great spheres $S_i, 1 \leq i \leq m$.

Now, let F be an arbitrary face of the convex body $\mathbf{K} \subset \mathbb{E}^d$, $d \geq 3$, and let B_F denote the smallest spherical ball of $\mathbb{S}^{d-1}$ with center $\mathbf{f} \in \mathbb{S}^{d-1}$ which contains the Gauss image $\nu(F)$ of F. By assumption the radius of B_F is at most r. As the family $\{B_i, 1 \leq i \leq m\}$ of antipodal pairs of balls forms a covering of $\mathbb{S}^{d-1}$ therefore $\mathbf{f} \in B_j$ for some $1 \leq j \leq m$. If, in addition, we have that $\mathbf{f} \in \operatorname{int} B_j$ (where int( ) denotes the interior of the corresponding set in $\mathbb{S}^{d-1}$), then the inequality $r + R \leq \frac{\pi}{2}$ implies that $\nu(F) \cap S_j = \emptyset$. If $\mathbf{f}$ does not belong to the interior of any of the sets $B_i, 1 \leq i \leq m$, then clearly $\mathbf{f}$ must be a boundary point of at least d sets of the family $\{B_i, 1 \leq i \leq m\}$. Then either we find an S_i being disjoint from $\nu(F)$ or we end up with d members of the family $\{S_i, 1 \leq i \leq m\}$ each being tangent to B_F at some point of $\nu(F)$. Clearly, the latter case can occur only for finitely many $\nu(F)$s and so, by taking a proper congruent copy of the great spheres $\{S_i, 1 \leq i \leq m\}$ within $\mathbb{S}^{d-1}$ (under which we mean to avoid finitely many so-called prohibited positions) we get that each $\nu(F)$ is disjoint from at least one member of the family $\{S_i, 1 \leq i \leq m\}$. This completes the proof of the first part of Theorem 3.5.2. Finally, the second part of Theorem 3.5.2 follows from the first one in a rather straightforward way.

9.3 Proof of Theorem 3.5.3 Using Antipodal Spherical Codes of Small Covering Radii

We show that any set of constant width in $\mathbb{E}^3$ can be X-rayed by 3 pairwise orthogonal directions (i.e., by 3 lines passing through the origin $\mathbf{o}$ of $\mathbb{E}^3$), one of which can be chosen arbitrarily. This is a somewhat stronger statement than the first claim of Theorem 3.5.3. Now, let us take the following special class of convex sets in $\mathbb{E}^3$. Let $Y \subset \mathbb{E}^3$ be an arbitrary set of diameter at most 1 and let $\mathbf{B}[Y]$ denote the intersection of the closed 3-dimensional unit balls centered at the points of Y. Using the fact that a closed set $Y \subset \mathbb{E}^3$ is of constant width 1 if, and only if, $\mathbf{B}[Y] = Y$ (cf. [125]), we obtain that the following theorem implies the statement at the beginning of this section and so, the first claim of Theorem 3.5.3.

Theorem 9.3.1 *Let $Y \subset \mathbb{E}^3$ be an arbitrary set of diameter at most 1. Then $\mathbf{B}[Y]$ can be X-rayed by 3 pairwise orthogonal directions of $\mathbb{E}^3$, one of which can be chosen arbitrarily.*

Proof: Let $\mathbf{K} \subset \mathbb{E}^3$ be a convex body and $\mathbf{b} \in \operatorname{bd}\mathbf{K}$. Then let $N_\mathbf{K}(\mathbf{b}) \subset \mathbb{S}^2$ denote the set of inward unit normal vectors of the planes that support $\mathbf{K}$ at $\mathbf{b}$. Moreover, if $L \subset \mathbb{E}^3$ is an arbitrary line passing through the origin $\mathbf{o}$ of $\mathbb{E}^3$, then let $C(L)$ denote the great circle of $\mathbb{S}^2$, whose plane is orthogonal to L. It is well known (see [72]) that L can X-ray the boundary point $\mathbf{b}$ of $\mathbf{K}$ if and only if $N_\mathbf{K}(\mathbf{b}) \cap C(L) = \emptyset$. We need the following lemma.

Lemma 9.3.2 *Let $Y \subset \mathbb{E}^3$ be an arbitrary set of diameter at most 1 and let $\mathbf{b} \in \mathrm{bd}\,(\mathbf{B}[Y])$. Then $N_{\mathbf{K}}(\mathbf{b}) \subset \mathbb{S}^2$ is of spherical diameter not greater than $\frac{\pi}{3}$.*

Proof: We may assume that Y is closed. It is sufficient to show that if $\mathbf{y}_1, \mathbf{y}_2 \in S^2(\mathbf{b}, 1) \cap Y$ (where $S^2(\mathbf{b}, 1)$ denotes the unit sphere centered at $\mathbf{b}$ in $\mathbb{E}^3$), then $\sphericalangle(\mathbf{y}_1 \mathbf{b} \mathbf{y}_2) \leq \frac{\pi}{3}$. Indeed, this is true, because the Euclidean isosceles triangle $\mathrm{conv}\{\mathbf{y}_1, \mathbf{b}, \mathbf{y}_2\}$ has two legs $\mathbf{y}_1 \mathbf{b}$ and $\mathbf{y}_2 \mathbf{b}$ of length 1, and a base, namely $\mathbf{y}_1 \mathbf{y}_2$ of length at most 1 due to the assumption that the diameter of Y is at most 1. This finishes the proof of Lemma 9.3.2.   □

Let $L_1 \subset \mathbb{E}^3$ be an arbitrary line passing through the origin $\mathbf{o}$ of $\mathbb{E}^3$. We call L_1 vertical, and lines perpendicular to L_1 horizontal. We pick two pairwise orthogonal, horizontal lines, L_2 and L_3. Assume that the lines L_1, L_2, and L_3 do not X-ray $\mathbf{B}[Y]$, where $Y \subset \mathbb{E}^3$ is an arbitrary set of diameter at most 1. Then there is a point $\mathbf{b} \in \mathrm{bd}\,(\mathbf{B}[Y])$ such that $N_{\mathbf{B}[Y]}(\mathbf{b}) \subset \mathbb{S}^2$ intersects each of the three great circles $C(L_1), C(L_2)$, and $C(L_3)$ of $\mathbb{S}^2$. We choose three points of $N_{\mathbf{B}[Y]}(\mathbf{b})$, one on each great circle: $\mathbf{z}_1 \in N_{\mathbf{B}[Y]}(\mathbf{b}) \cap C(L_1), \mathbf{z}_2 \in N_{\mathbf{B}[Y]}(\mathbf{b}) \cap C(L_2)$, and $\mathbf{z}_3 \in N_{\mathbf{B}[Y]}(\mathbf{b}) \cap C(L_3)$. Note that each of the three great circles is dissected into four equal arcs (of length $\frac{\pi}{4}$) by the two other great circles. By Lemma 9.3.2, $N_{\mathbf{B}[Y]}(\mathbf{b}) \subset \mathbb{S}^2$ is a spherically convex set of spherical diameter at most $\frac{\pi}{3}$. However, $\mathbf{z}_1, \mathbf{z}_2, \mathbf{z}_3 \in N_{\mathbf{B}[Y]}(\mathbf{b})$, so the generalization of Jung's theorem for spherical space by Dekster [119] shows that $\mathbf{z}_1, \mathbf{z}_2$, and $\mathbf{z}_3$ are the midpoints of the great circular arcs mentioned above. So, the only way that the point $\mathbf{b} \in \mathrm{bd}\,(\mathbf{B}[Y])$ is not X-rayed by any of the lines L_1, L_2, and L_3 is the following. The set $N_{\mathbf{B}[Y]}(\mathbf{b})$ contains a spherical equilateral triangle of spherical side length $\frac{\pi}{3}$ and the vertices of this spherical triangle lie on $C(L_1), C(L_2)$, and $C(L_3)$, respectively. Furthermore, each vertex is necessarily the midpoint of the quarter arc of the great circle on which it lies, and $N_{\mathbf{B}[Y]}(\mathbf{b})$ does not intersect either of the three great circles in any other point. Because the set $\{N_{\mathbf{B}[Y]}(\mathbf{b}') : \mathbf{b}' \in \mathrm{bd}\,(\mathbf{B}[Y])\}$ is a tiling of $\mathbb{S}^2$, there are only finitely many boundary points $\mathbf{b}' \in \mathrm{bd}\,(\mathbf{B}[Y])\}$ such that $N_{\mathbf{B}[Y]}(\mathbf{b}')$ contains an equilateral triangle of side length $\frac{\pi}{3}$ with a vertex on $C(L_1)$. We call these tiles blocking tiles. Now, by rotating L_2 and L_3 together in the horizontal plane, we can easily avoid all the blocking tiles; that is, we can find a rotation R about the line L_1 such that none of the blocking tiles has a vertex on both circles $C(R(L_2))$ and $C(R(L_3))$. Now, $L_1, R(L_2)$, and $R(L_3)$ are the desired directions finishing the proof of Theorem 9.3.1.   □

Now, we turn to a proof of the case $d = 4$ of Theorem 3.5.3. Based on Theorem 3.5.2, it is sufficient to find 12 pairwise antipodal points of $\mathbb{S}^3$ whose covering radius is at most $\alpha = \pi/2 - r_3 = \pi/2 - \arccos\sqrt{\frac{5}{8}}$. In order to achieve this let us take two regular hexagons of edge length 1 inscribed into $\mathbb{S}^3$ such that their 2-dimensional planes are totally orthogonal to each other in $\mathbb{E}^4$. Now, let $\mathbf{P}$ be the convex hull of the 12 vertices of the two regular hexagons. If F is any facet of $\mathbf{P}$, then it is easy to see that F is a 3-dimensional simplex having two pairs of vertices belonging to different hexagons with the property that each pair is in fact, a pair of two consecutive vertices of the relevant hexagon. As an obvious corollary of this we get that if one projects any facet of $\mathbf{P}$ from the center $\mathbf{o}$ of $\mathbb{S}^3$ onto $\mathbb{S}^3$, then the projection is a 3-dimensional spherical simplex whose two opposite edges are of length $\frac{\pi}{3}$ and the other four remaining edges are of length $\frac{\pi}{2}$. Also, it is easy to show that the circumradius of that spherical simplex is equal to $\alpha = \arccos\sqrt{\frac{3}{8}}$. This means that the covering radius of the 12 points in question lying in $\mathbb{S}^3$ is precisely α, finishing the proof of Theorem 3.5.3 for $d = 4$.

In dimensions $d = 5, 6$ we proceed similarly using Theorem 3.5.2. More exactly, we are going to construct 2^d pairwise antipodal points on $\mathbb{S}^{d-1}$ ($d = 5, 6$) with covering radius at most $\pi/2 - \arccos\sqrt{\frac{d+1}{2d}} = \arccos\sqrt{\frac{d-1}{2d}}$.

For $d = 5$ we need to find 32 pairwise antipodal points on $\mathbb{S}^4$ with covering radius at most $\arccos\sqrt{\frac{2}{5}} = 50.768...°$. Let us take a 2-dimensional plane $\mathbb{E}^2$ and a 3-dimensional subspace $\mathbb{E}^3$ in $\mathbb{E}^5$ such that they are totally orthogonal to each other (with both passing through the origin $\mathbf{o}$ of $\mathbb{E}^5$). Let $\mathbf{P}_2$ be a regular 16-gon inscribed into $\mathbb{E}^2 \cap \mathbb{S}^4$ and let $\mathbf{P}_3$ be a set of 16 pairwise antipodal points on $\mathbb{E}^3 \cap \mathbb{S}^4$ with covering radius $R_c = 33.547...°$. For the details of the construction of $\mathbf{P}_3$ see [129]. Finally, let $\mathbf{P}$ be the convex hull of $\mathbf{P}_2 \cup \mathbf{P}_3$. If F is any facet of $\mathbf{P}$, then it is easy to see that F is a 4-dimensional simplex having two vertices in $\mathbf{P}_2$ and three vertices in $\mathbf{P}_3$. If one projects F from the center $\mathbf{o}$ of $\mathbb{S}^4$ onto $\mathbb{S}^4$, then the projection F' is a 4-dimensional spherical symplex. Among its five vertices there are two vertices say, $\mathbf{a}$ and $\mathbf{b}$ lying in $\mathbf{P}_2$. Here $\mathbf{a}$ and $\mathbf{b}$ must be consecutive vertices of the regular 16-gon inscribed into $\mathbb{E}^2 \cap \mathbb{S}^4$, and the remaining three vertices must form a triangle inscribed into $\mathbb{E}^3 \cap \mathbb{S}^4$ with circumscribed circle $\mathcal{C}$ of radius R_c. Now, let $\mathbf{c}'$ be the center of $\mathcal{C}$ and $\mathbf{c}$ be an arbitrary point of $\mathcal{C}$; moreover, let $\mathbf{m}$ be the midpoint of spherical segment $\mathbf{ab}$. Clearly, $\mathbf{am} = 11.25°$, $\mathbf{c'm} = 90°$, and $\mathbf{cc'} = R_c$ on $\mathbb{S}^4$. If $\mathbf{s}$ denotes the center of the circumscribed sphere of F' in $\mathbb{S}^4$, then $\mathbf{s}$ is a point of the spherical segment $\mathbf{c'm}$. Let $\mathbf{as} = \mathbf{bs} = \mathbf{cs} = x$, $\mathbf{sm} = y$ and $\mathbf{c's} = 90° - y$. Now, the cosine theorem applied to the spherical right triangles $\varDelta\mathbf{ams}$ and $\varDelta\mathbf{cc's}$ implies that

$$\cos x = \cos 11.25° \cdot \cos y, \text{ and } \cos x = \cos R_c \sin y.$$

By solving these equations for x and y we get that $x = 50.572...° < \arccos\sqrt{\frac{2}{5}}$ $= 50.768...°$ finishing the proof of Theorem 3.5.3 for $d = 5$.

For $d = 6$ we need to construct 64 pairwise antipodal points on $\mathbb{S}^5$ with covering radius at most $\arccos\sqrt{\frac{5}{12}} = 49.797...°$. In order to achieve this let us take two 3-dimensional subspaces $\mathbb{E}_1^3$ and $\mathbb{E}_2^3$ in $\mathbb{E}^6$ such that they are totally orthogonal to each other (with both passing through the origin $\mathbf{o}$ of $\mathbb{E}^6$). For $i = 1, 2$ let $\mathbf{P}_i$ be a set of 32 pairwise antipodal points on $\mathbb{E}_i^3 \cap \mathbb{S}^5$ with covering radius $R_c = 22.690...°$. For the details of the construction of $\mathbf{P}_i, i = 1, 2$ see [129]. Finally, let $\mathbf{P}$ be the convex hull of $\mathbf{P}_1 \cup \mathbf{P}_2$. If F is any facet of $\mathbf{P}$, then it is easy to see that F is a 5-dimensional simplex having three vertices both in $\mathbf{P}_1$ and in $\mathbf{P}_2$. If one projects F from the center $\mathbf{o}$ of $\mathbb{S}^5$ onto $\mathbb{S}^5$, then the projection F' is a 5-dimensional spherical symplex. It follows from the construction above that two spherical triangles formed by the two proper triplets of the vertices of F' have circumscribed circles $\mathcal{C}_1$ and $\mathcal{C}_2$ of radius R_c. If $\mathbf{c}_i$ denotes the center of $\mathcal{C}_i, i = 1, 2$ and $\mathbf{s}$ denotes the center of the circumscribed sphere of F' in $\mathbb{S}^5$, then it is easy to show that $\mathbf{s}$ is, in fact, the midpoint of the spherical segment $\mathbf{c}_1\mathbf{c}_2$ whose spherical length is of $90°$. Thus, if x denotes the spherical radius of the circumscribed sphere of F' in $\mathbb{S}^5$, then the cosine theorem applied to the proper spherical right triangle implies that

$$\cos x = \cos R_c \cdot \cos 45°.$$

Hence, it follows that $x = 49.278...° < \arccos \sqrt{\frac{5}{12}} = 49.797...°$ finishing the proof of Theorem 3.5.3 for $d = 6$.

9.4 Proofs of Theorem 3.8.1 and Theorem 3.8.3

9.4.1 From the Banach–Mazur distance to the vertex index

We identify the d-dimensional affine space with $\mathbb{R}^d$. By $|\cdot|$ and $\langle \cdot, \cdot \rangle$ we denote the canonical Euclidean norm and the canonical inner product on $\mathbb{R}^d$. The canonical basis of $\mathbb{R}^d$ we denote by $\mathbf{e}_1, \ldots, \mathbf{e}_d$. By $\|\cdot\|_p, 1 \leq p \leq \infty$, we denote the ℓ_p-norm; that is,

$$\|\mathbf{x}\|_p := \left(\sum_{1 \leq i \leq d} |x_i|^p \right)^{1/p} \quad \text{for } p < \infty \quad \text{and} \quad \|\mathbf{x}\|_\infty := \sup\{|x_i| \mid 1 \leq i \leq d\}$$

where $\mathbf{x} = (x_1, \ldots, x_d)$. In particular, $\|\cdot\|_2 = |\cdot|$. As usual, $\ell_p^d = (\mathbb{R}^d, \|\cdot\|_p)$, and the unit ball of ℓ_p^d is denoted by $\mathbf{B}_p^d$.

Given points $\mathbf{x}_1, \ldots, \mathbf{x}_k$ in $\mathbb{R}^d$ we denote their convex hull by $\text{conv}\{\mathbf{x}_i\}_{i \leq k}$ and their absolute convex hull by $\text{absconv}\{\mathbf{x}_i\}_{i \leq k} := \text{conv}\{\pm\mathbf{x}_i\}_{i \leq k}$. Similarly, the convex hull of a set $A \subset \mathbb{R}^d$ is denoted by $\text{conv} A$ and the absolute convex hull of A is denoted by $\text{absconv} A := \text{conv}(A \cup -A)$.

Let $\mathbf{K} \subset \mathbb{R}^d$ be a convex body, that is, a compact convex set with non-empty interior such that the origin $\mathbf{o}$ of $\mathbb{R}^d$ belongs to $\mathbf{K}$. We denote by $\mathbf{K}^\circ$ the polar of $\mathbf{K}$, that is,

$$\mathbf{K}^\circ := \{\mathbf{x} \mid \langle \mathbf{x}, \mathbf{y} \rangle \leq 1 \text{ for every } \mathbf{y} \in \mathbf{K}\}.$$

If $\mathbf{K}$ is an $\mathbf{o}$-symmetric convex body, then the Minkowski functional of $\mathbf{K}$,

$$\|\mathbf{x}\|_{\mathbf{K}} := \inf\{\lambda > 0 \mid \mathbf{x} \in \lambda\mathbf{K}\},$$

defines a norm on $\mathbb{R}^d$ with the unit ball $\mathbf{K}$.

The *Banach–Mazur distance* between two **o**-symmetric convex bodies $\mathbf{K}$ and $\mathbf{L}$ in $\mathbb{R}^d$ is defined by

$$d(\mathbf{K}, \mathbf{L}) := \inf \{\lambda > 0 \mid \mathbf{L} \subset T\mathbf{K} \subset \lambda \mathbf{L}\},$$

where the infimum is taken over all linear operators $T : \mathbb{R}^d \to \mathbb{R}^d$. It is easy to see that

$$d(\mathbf{K}, \mathbf{L}) = d(\mathbf{K}^\circ, \mathbf{L}^\circ).$$

The Banach–Mazur distance between $\mathbf{K}$ and the closed Euclidean ball $\mathbf{B}_2^d$ we denote by $d_{\mathbf{K}}$. As is well known, John's theorem ([172]) implies that for every **o**-symmetric convex body $\mathbf{K}$, $d_{\mathbf{K}}$ is bounded by $\sqrt{d}$.

Given a (convex) body $\mathbf{K}$ in $\mathbb{R}^d$ we denote its volume by $\mathrm{vol}_d(\mathbf{K})$. Let $\mathbf{K}$ be an **o**-symmetric convex body in $\mathbb{R}^d$. The *outer volume ratio* of $\mathbf{K}$ is

$$\mathrm{ovr}(\mathbf{K}) := \inf \left(\frac{\mathrm{vol}_d(\mathcal{E})}{\mathrm{vol}_d(\mathbf{K})} \right)^{1/d},$$

where the infimum is taken over all **o**-symmetric ellipsoids $\mathcal{E}$ in $\mathbb{R}^d$ containing $\mathbf{K}$. By John's theorem we have

$$\mathrm{ovr}(\mathbf{K}) \leq \sqrt{d}.$$

Lemma 9.4.1 *Let $\mathbf{K}$ and $\mathbf{L}$ be **o**-symmetric convex bodies in $\mathbb{R}^d$. Then*

$$\mathrm{vein}(\mathbf{K}) \leq d(\mathbf{K}, \mathbf{L}) \cdot \mathrm{vein}(\mathbf{L}).$$

Proof: Let T be a linear operator such that $\mathbf{K} \subset T\mathbf{L} \subset \lambda \mathbf{K}$. Let $\mathbf{p}_1, \mathbf{p}_2, \ldots, \mathbf{p}_n \in \mathbb{R}^d$ be such that $\mathrm{conv}\{\mathbf{p}_i\}_{1 \leq i \leq n} \supset \mathbf{L}$. Then $\mathrm{conv}\{T\mathbf{p}_i\}_{1 \leq i \leq n} \supset T\mathbf{L} \supset \mathbf{K}$. Because $T\mathbf{L} \subset \lambda \mathbf{K}$, we also have $\|\cdot\|_{\mathbf{K}} \leq \lambda \|\cdot\|_{T\mathbf{L}}$. Therefore,

$$\sum_{1 \leq i \leq n} \|T\mathbf{p}_i\|_{\mathbf{K}} \leq \lambda \sum_{1 \leq i \leq n} \|T\mathbf{p}_i\|_{T\mathbf{L}} = \lambda \sum_{1 \leq i \leq n} \|\mathbf{p}_i\|_{\mathbf{L}},$$

which implies the desired result. $\qquad\qquad\square$

9.4.2 Calculating the vertex index of Euclidean balls in dimensions 2 and 3

Theorem 9.4.2
(i) For the Euclidean balls in $\mathbb{R}^2$ and $\mathbb{R}^3$ we have

$$\mathrm{vein}(\mathbf{B}_2^2) = 4\sqrt{2}, \qquad \mathrm{vein}(\mathbf{B}_2^3) = 6\sqrt{3}.$$

*(ii) In general, if $\mathbf{K} \subset \mathbb{R}^2$, $\mathbf{L} \subset \mathbb{R}^3$ are arbitrary **o**-symmetric convex bodies, then*

$$4 \leq \mathrm{vein}(\mathbf{K}) \leq 6 \leq \mathrm{vein}(\mathbf{L}) \leq 18.$$

Proof: We prove (i) as follows. As $\mathbf{B}_2^d \subset \sqrt{d}\,\mathbf{B}_1^d$ for all d, therefore $\mathrm{vein}(\mathbf{B}_2^d) \le 2d\sqrt{d}$. So, we need to prove the lower estimates $\mathrm{vein}(\mathbf{B}_2^2) \ge 4\sqrt{2}$ and $\mathrm{vein}(\mathbf{B}_2^3) \ge 6\sqrt{3}$.

Let $\mathbf{P} \subset \mathbb{R}^2$ be a convex polygon with vertices $\mathbf{p}_1, \mathbf{p}_2, \ldots, \mathbf{p}_n, n \ge 3$ containing $\mathbf{B}_2^2$. Let $\mathbf{P}^\circ$ denote the polar of $\mathbf{P}$. Assume that the side of $\mathbf{P}^\circ$ corresponding to the vertex $\mathbf{p}_i$ of $\mathbf{P}$ generates the central angle $2\alpha_i$ with vertex $\mathbf{o}$. Clearly, $0 < \alpha_i < \pi/2$ and $|\mathbf{p}_i| \ge \frac{1}{\cos \alpha_i}$ for all $i \le n$. As $\frac{1}{\cos x}$ is a convex function over the open interval $(-\pi/2, \pi/2)$ therefore the Jensen inequality implies that

$$\sum_{i=1}^n |\mathbf{p}_i| \ge \sum_{i=1}^n \frac{1}{\cos \alpha_i} \ge \frac{n}{\cos\left(\frac{\sum_{i=1}^n \alpha_i}{n}\right)} = \frac{n}{\cos \frac{\pi}{n}}.$$

It is easy to see that $\frac{n}{\cos(\pi/n)} \ge \frac{4}{\cos(\pi/4)} = 4\sqrt{2}$ holds for all $n \ge 3$. Thus, $\mathrm{vein}(\mathbf{B}_2^2) \ge 4\sqrt{2}$. This completes the proof of the planar case.

Now, we handle the 3-dimensional case. Let $\mathbf{P} \subset \mathbb{R}^3$ be a convex polyhedron with vertices $\mathbf{p}_1, \mathbf{p}_2, \ldots, \mathbf{p}_n, n \ge 4$, containing $\mathbf{B}_2^3$. Of course, we assume that $|\mathbf{p}_i| > 1$. We distinguish the following three cases: (a) $n = 4$, (b) $n \ge 8$, and (c) $5 \le n \le 7$. In fact, the proof given for Case (c) also works for Case (b), however, Case (b) is much simpler, so we have decided to consider it separately.

Case (a): $n = 4$. In this case $\mathbf{P}$ is a tetrahedron with triangular faces T_1, T_2, T_3, and T_4. Without loss of generality we may assume that $\mathbf{B}_2^3$ is tangential to the faces T_1, T_2, T_3, and T_4. Then the well-known inequality between the harmonic and arithmetic means yields that

$$1 = \sum_{i=1}^4 \frac{\frac{1}{3}\mathrm{vol}_2(T_i)}{\mathrm{vol}_3(\mathbf{P})} \ge \sum_{i=1}^4 \frac{1}{|\mathbf{p}_i| + 1} \ge \frac{4^2}{\sum_{i=1}^4 (|\mathbf{p}_i| + 1)}.$$

This implies in a straightforward way that

$$\sum_{i=1}^4 |\mathbf{p}_i| \ge 12 > 6\sqrt{3},$$

finishing the proof of this case.

For the next two cases we need the following notation. Fix $i \le n$. Let C_i denote the (closed) spherical cap of $\mathbb{S}^2$ with spherical radius R_i which is the union of points $x \in \mathbb{S}^2$ such that the open line segment connecting $\mathbf{x}$ and $\mathbf{p}_i$ is disjoint from $\mathbf{B}_2^3$. In other words, C_i is the (closed) spherical cap with the center $\mathbf{p}_i/|\mathbf{p}_i|$ and the spherical radius R_i, satisfying $|\mathbf{p}_i| = \frac{1}{\cos R_i}$. By b_i we denote the spherical area of C_i. Then $b_i = 2\pi(1 - \cos R_i)$.

Case (b): $n \ge 8$. Because $\mathbf{P}$ contains $\mathbf{B}_2^3$, we have

$$\mathbb{S}^2 \subset \bigcup_{i=1}^n C_i.$$

Comparing the areas, we observe

$$4\pi \leq \sum_{i=1}^{n} b_i = \sum_{i=1}^{n} 2\pi \left(1 - \cos R_i\right),$$

which implies

$$\sum_{i=1}^{n} \cos R_i \leq n - 2.$$

Applying again the inequality between the harmonic and arithmetic means, we obtain

$$\sum_{i=1}^{n} |\mathbf{p}_i| = \sum_{i=1}^{n} \frac{1}{\cos R_i} \geq \frac{n^2}{\sum_{i=1}^{n} \cos R_i} \geq \frac{n^2}{n-2} \geq \frac{64}{6} > 6\sqrt{3}.$$

Case (c): $5 \leq n \leq 7$. Let $\mathbf{P}^\circ$ denote the polar of $\mathbf{P}$. Given $i \leq n$, let F_i denote the central projection of the face of $\mathbf{P}^\circ$ that corresponds to the vertex $\mathbf{p}_i$ of $\mathbf{P}$ from the center $\mathbf{o}$ onto the boundary of $\mathbf{B}_2^3$, that is, onto the unit sphere $\mathbb{S}^2$ centered at $\mathbf{o}$. Obviously, F_i is a spherically convex polygon of $\mathbb{S}^2$ and $F_i \subset C_i$. Let n_i denote the number of sides of F_i and let a_i stand for the spherical area of F_i. Note that the area of the sphere is equal to the sum of areas of F_is; that is $\sum_{i=1}^{n} a_i = 4\pi$. As $10 < 6\sqrt{3} = 10.3923... < 11$, therefore without loss of generality we may assume that there is no i for which $|\mathbf{p}_i| = \frac{1}{\cos R_i} \geq 11 - 3 = 8$; in other words we assume that $0 < R_i <$ arccos $\frac{1}{8} = 1.4454... < \frac{\pi}{2}$ for all $i \leq n$. Note that this immediately implies that $0 < a_i < b_i = 2\pi(1 - \cos R_i) < \frac{7\pi}{4} < 5.5$ for all $1 \leq i \leq n$.

It is well known that if $C \subset \mathbb{S}^2$ is a (closed) spherical cap of radius less than $\frac{\pi}{2}$, then the spherical area of a spherically convex polygon with at most $s \geq 3$ sides lying in C is maximal for the regular spherically convex polygon with s sides inscribed in C. (This can be easily obtained with the help of the Lexell circle (see [134]).) It is also well known that if F_i^* denotes a regular spherically convex polygon with n_i sides and of spherical area a_i, and if R_i^* denotes the circumradius of F_i^*, then $\frac{1}{\cos R_i^*} = \tan \frac{\pi}{n_i} \tan \left(\frac{a_i + (n_i - 2)\pi}{2n_i}\right)$. Thus, for every $i \leq n$ we have

$$|\mathbf{p}_i| = \frac{1}{\cos R_i} \geq \tan \frac{\pi}{n_i} \tan \left(\frac{a_i + (n_i - 2)\pi}{2n_i}\right).$$

Here $3 \leq n_i \leq n - 1 \leq 6$ and $0 < a_i < \frac{7\pi}{4}$ for all $1 \leq i \leq n$.

Now, it is natural to consider the function $f(x,y) = \tan \frac{\pi}{y} \tan \left(\frac{x + (y-2)\pi}{2y}\right)$ defined on $\{(x,y) \mid 0 < x < 2\pi,\ 3 \leq y\}$. As in the 2-dimensional case we use the Jensen inequality. But, unfortunately, it turns out that f is convex only on a proper subset of its domain; namely the following holds. (This can be proved using a standard analytic approach or tools such as MAPLE. We omit the details.)

Lemma 9.4.3 *Let f be a function of two variables defined by*

$$f(x, y) = \tan \frac{\pi}{y} \tan \left(\frac{x + (y-2)\pi}{2y} \right).$$

Then
(i) for every fixed $0 < x_0 < 2\pi$ the function $f(x_0, y)$ is decreasing in y over the interval $[3, \infty)$,
(ii) for every fixed $y_0 \geq 3$ the function $f(x, y_0)$ is increasing in x over the interval $(0, 2\pi)$,
(iii) for every fixed $y_0 \geq 3$ the function $f(x, y_0)$ is convex on the interval $(0, 2\pi)$,
(iv) f is convex on the closed rectangle $\{(x, y) \mid 0.4 \leq x \leq 5.5, 3 \leq y \leq 9\}$.

Without loss of generality we may assume that m is chosen such that $0 < a_i < 0.4$ for all $i \leq m$ and $0.4 \leq a_i < 5.5$ for all $m+1 \leq i \leq n$. Because $\sum_{i=1}^{n} a_i = 4\pi$, one has $m < n - 1$. By Lemma 9.4.3 (iv) and by the Jensen inequality, we obtain

$$\sum_{i=1}^{n} |\mathbf{p}_i| \geq \sum_{i=1}^{m} |\mathbf{p}_i| + \sum_{i=m+1}^{n} f(a_i, \, n_i)$$

$$\geq m + (n - m) \, f \left(\frac{1}{n - m} \sum_{i=m+1}^{n} a_i, \; \frac{1}{n - m} \sum_{i=m+1}^{n} n_i \right)$$

(here by $\sum_{i}^{0}$ we mean 0). Because $\sum_{i=1}^{n} a_i = 4\pi$, we have $\sum_{i=m+1}^{n} a_i > 4\pi - 0.4m$. By Euler's theorem on the edge graph of $\mathbf{P}^\circ$ we also have that $\sum_{i=1}^{n} n_i \leq 6n - 12$ and therefore $\sum_{i=m+1}^{n} n_i \leq (6n - 12) - 3m$. Thus, applying Lemma 9.4.3 (i) and (ii), we observe

$$\sum_{i=1}^{n} |\mathbf{p}_i| \geq m + (n - m)f \left(\frac{4\pi - 0.4m}{n - m}, \; \frac{(6n - 12) - 3m}{n - m} \right) =: g(m, n).$$

First we show that $g(m, n) \geq 6\sqrt{3} = 10.3923...$ for every (m, n) with $6 \leq n \leq 7$ and $0 \leq m < n - 1$.
Subcase $n = 7$:

$$g(0, 7) = 10.9168..., \quad g(1, 7) = 10.8422..., \quad g(2, 7) = 10.8426...,$$

$$g(3, 7) = 11.0201..., \quad g(4, 7) = 11.7828..., \quad g(5, 7) = 18.3370....$$

Subcase $n = 6$:

$$g(0, 6) = 6\sqrt{3} = 10.3923..., \quad g(1, 6) = 10.4034..., \quad g(2, 6) = 10.6206...,$$

$$g(3, 6) = 11.5561..., \quad g(4, 6) = 21.2948....$$

Subcase $n = 5$: First note that

$$6\sqrt{3} < g(1,5) = 10.6302... < g(2,5) = 11.8680... < g(3,5) = 28.1356....$$

Unfortunately, $g(0,5) < 6\sqrt{3}$, so we treat the case $n = 5$ slightly differently (in fact the proof is easier than the proof of the case $6 \leq n \leq 7$, because we use convexity of a function of one variable). In this case $\mathbf{P}$ has only 5 vertices, so it is either a double tetrahedron or a cone over a quadrilateral. As the latter one can be thought of as a limiting case of double tetrahedra, we can assume that the edge graph of $\mathbf{P}$ has two vertices, say $\mathbf{p}_1$ and $\mathbf{p}_2$, of degree three and three vertices, say $\mathbf{p}_3, \mathbf{p}_4$, and $\mathbf{p}_5$, of degree four. Thus $n_1 = n_2 = 3$ and $n_3 = n_4 = n_5 = 4$. Therefore

$$\sum_{i=1}^{5} |\mathbf{p}_i| \geq \sum_{i=1}^{5} f(a_i, \ n_i) = \sum_{i=1}^{2} f(a_i, \ 3) + \sum_{i=3}^{5} f(a_i, \ 4).$$

By Lemma 9.4.3 (iii) and by the Jensen inequality, we get

$$\sum_{i=1}^{5} |\mathbf{p}_i| \geq 2 \ f\left(\frac{a_1 + a_2}{2}, \ 3\right) + 3 \ f\left(\frac{a_3 + a_4 + a_5}{3}, \ 4\right)$$

$$= 2 \ f(a, \ 3) + 3 \ f\left(\frac{4\pi - 2a}{3}, \ 4\right)$$

$$= 2\sqrt{3} \ \tan\left(\frac{a + \pi}{6}\right) + 3 \ \tan\left(\frac{5\pi - a}{12}\right) =: h(a),$$

where $0 \leq a = \frac{a_1 + a_2}{2} < 5.5$. Finally, it is easy to show that the minimum value of $h(a)$ over the closed interval $0 \leq a \leq 5.5$ is (equal to 10.5618... and therefore is) strictly larger than $6\sqrt{3} = 10.3923...$, completing the proof of part (i) of Theorem 9.4.2.

Finally, we prove part (ii) of Theorem 9.4.2 as follows. First, observe that (i), John's theorem, and Lemma 9.4.1 imply that

$$4 = \frac{4\sqrt{2}}{\sqrt{2}} \leq \frac{\text{vein}(\mathbf{B}_2^2)}{d_{\mathbf{K}}} \leq \text{vein}(\mathbf{K})$$

and

$$6 = \frac{6\sqrt{3}}{\sqrt{3}} \leq \frac{\text{vein}(\mathbf{B}_2^3)}{d_{\mathbf{L}}} \leq \text{vein}(\mathbf{L}) \leq d_{\mathbf{L}} \cdot \text{vein}(\mathbf{B}_2^3) \leq 18.$$

Second, Theorem 3.7.1 implies that indeed $\text{vein}(\mathbf{K}) \leq 6$, finishing the proof of Theorem 9.4.2.   □

9.4.3 A lower bound for the vertex index using the Blaschke–Santaló inequality and an inequality of Ball and Pajor

Theorem 9.4.4 *For every $d \geq 2$ and every $\mathbf{o}$-symmetric convex body $\mathbf{K}$ in $\mathbb{R}^d$ one has*

$$\frac{d^{3/2}}{\sqrt{2\pi e}\,\, \mathrm{ovr}(\mathbf{K})} \leq \frac{d}{\left(\mathrm{vol}_d(\mathbf{B}_2^d)\right)^{1/d}\,\mathrm{ovr}(\mathbf{K})} \leq \mathrm{vein}(\mathbf{K}).$$

Proof: Recall that $\mathrm{vein}(\mathbf{K})$ is an affine invariant; that is, $\mathrm{vein}(\mathbf{K}) = \mathrm{vein}(T\mathbf{K})$ for every invertible linear operator $T : \mathbb{R}^d \to \mathbb{R}^d$. Thus, without loss of generality we can assume that $\mathbf{B}_2^d$ is the ellipsoid of minimal volume for $\mathbf{K}$. In particular, $\mathbf{K} \subset \mathbf{B}_2^d$, so $|\cdot| \leq \|\cdot\|_{\mathbf{K}}$.

Let $\{\mathbf{p}_i\}_1^N \in \mathbb{R}^d$ be such that $\mathbf{K} \subset \mathrm{conv}\{\mathbf{p}_i\}_1^N$. Clearly $N \geq d+1$. Denote

$$\mathbf{L} := \mathrm{absconv}\{\mathbf{p}_i\}_1^N.$$

Then

$$\mathbf{L}^\circ = \{\mathbf{x} \mid |\langle \mathbf{x}, \mathbf{p}_i \rangle| \leq 1 \text{ for every } i \leq N\}.$$

By Theorem 2 of [11], we get that

$$\mathrm{vol}_d\left(\mathbf{L}^\circ\right) \geq \left(\frac{d}{\sum_1^N |\mathbf{p}_i|}\right)^d.$$

On the one hand, according to the Blaschke–Santaló inequality (see, e.g., [215]) $\mathrm{vol}_d\left(\mathbf{L}\right)\mathrm{vol}_d\left(\mathbf{L}^\circ\right) \leq \left(\mathrm{vol}_d\left(\mathbf{B}_2^d\right)\right)^2$. On the other hand, $\mathbf{K} \subset \mathbf{L}$, therefore the above inequality implies that

$$\mathrm{vol}_d\left(\mathbf{K}\right) \leq \mathrm{vol}_d\left(\mathbf{L}\right) \leq \frac{\left(\mathrm{vol}_d\left(\mathbf{B}_2^d\right)\right)^2}{\mathrm{vol}_d\left(\mathbf{L}^\circ\right)} \leq \left(\mathrm{vol}_d\left(\mathbf{B}_2^d\right)\right)^2 \left(\frac{1}{d}\sum_1^N |\mathbf{p}_i|\right)^d.$$

As $\mathbf{B}_2^d$ is the minimal volume ellipsoid of $\mathbf{K}$ and as $\|\cdot\|_{\mathbf{K}} \geq |\cdot|$, therefore we conclude that

$$\frac{1}{\mathrm{ovr}(\mathbf{K})} = \left(\frac{\mathrm{vol}_d\left(\mathbf{K}\right)}{\mathrm{vol}_d\left(\mathbf{B}_2^d\right)}\right)^{1/d} \leq \left(\mathrm{vol}_d\left(\mathbf{B}_2^d\right)\right)^{1/d}\frac{1}{d}\sum_1^N \|\mathbf{p}_i\|_{\mathbf{K}}.$$

Thus, the inequality

$$\frac{d}{\left(\mathrm{vol}_d(\mathbf{B}_2^d)\right)^{1/d}\,\mathrm{ovr}(\mathbf{K})} \leq \sum_1^N \|\mathbf{p}_i\|_{\mathbf{K}}$$

together with the well-known inequality (see, e.g., [215])

$$\text{vol}_d(\mathbf{B}_2^d) = \frac{\pi^{d/2}}{\Gamma(1+d/2)} \leq \left(\frac{2\pi e}{d}\right)^{d/2}$$

finishes the proof of Theorem 9.4.4.   □

As an immediate corollary of Theorem 9.4.4 we get the lower bound in the following statement. (The upper bound mentioned there has been derived earlier as a rather simple observation.)

Corollary 9.4.5 *For every $d \geq 2$ one has*

$$\frac{d^{3/2}}{\sqrt{2\pi e}} \leq \text{vein}(\mathbf{B}_2^d) \leq 2\ d^{3/2}.$$

9.4.4 An upper bound for the vertex index using a theorem of Rudelson

Let $\mathbf{u}, \mathbf{v} \in \mathbb{R}^d$. As usual $Id : \mathbb{R}^d \to \mathbb{R}^d$ denotes the identity operator and $\mathbf{u} \otimes \mathbf{v}$ denotes the operator from $\mathbb{R}^d$ to $\mathbb{R}^d$, defined by $(\mathbf{u} \otimes \mathbf{v})(\mathbf{x}) = \langle \mathbf{u}, \mathbf{x} \rangle \mathbf{v}$ for every $\mathbf{x} \in \mathbb{R}^d$. In [223] and [224], M. Rudelson proved the following theorem (see Corollary 4.3 of [224] and Theorem 1.1 with Remark 4.1 of [223]).

Theorem 9.4.6 *For every $\mathbf{o}$-symmetric convex body $\mathbf{K}$ in $\mathbb{R}^d$ and every $\epsilon \in (0, 1]$ there exists an $\mathbf{o}$-symmetric convex body $\mathbf{L}$ in $\mathbb{R}^d$ such that $d(\mathbf{K}, \mathbf{L}) \leq 1 + \epsilon$ and $\mathbf{B}_2^d$ is the minimal volume ellipsoid containing $\mathbf{L}$, and*

$$Id = \sum_{i=1}^{M} c_i \mathbf{u}_i \otimes \mathbf{u}_i,$$

where $c_1, \ldots, c_M$ are positive numbers, $\mathbf{u}_1, \ldots, \mathbf{u}_M$ are contact points of $\mathbf{L}$ and $\mathbf{B}_2^d$ (i.e., $\|\mathbf{u}_i\|_{\mathbf{L}} = |\mathbf{u}_i| = 1$), and

$$M \leq C\ \epsilon^{-2}\ d\ \ln(2d),$$

with an absolute constant C.

It is a standard observation (cf. [14], [243]) that under the conditions of Theorem 9.4.6 one has
$$\mathbf{P} \subset \mathbf{L} \subset \mathbf{B}_2^d \subset \sqrt{d}\ \mathbf{L},$$

for $\mathbf{P} = \text{absconv}\{\mathbf{u}_i\}_{i \leq M}$. Indeed, $\mathbf{P} \subset \mathbf{L}$ by the convexity and the symmetry of $\mathbf{L}$, and for every $\mathbf{x} \in \mathbb{R}^d$ we have

$$\mathbf{x} = Id\ \mathbf{x} = \sum_{i=1}^{M} c_i \langle \mathbf{u}_i, \mathbf{x} \rangle \mathbf{u}_i,$$

so

$$|\mathbf{x}|^2 = \langle \mathbf{x}, \mathbf{x} \rangle = \sum_{i=1}^{M} c_i \langle \mathbf{u}_i, \mathbf{x} \rangle^2 \leq \max_{i \leq M} \langle \mathbf{u}_i, \mathbf{x} \rangle^2 \sum_{i=1}^{M} c_i = \|\mathbf{x}\|_{\mathbf{P}^\circ}^2 \sum_{i=1}^{M} c_i.$$

Because

$$d = \operatorname{trace} Id = \operatorname{trace} \sum_{i=1}^{M} c_i \mathbf{u}_i \otimes \mathbf{u}_i = \sum_{i=1}^{M} c_i \langle \mathbf{u}_i, \mathbf{u}_i \rangle = \sum_{i=1}^{M} c_i,$$

we obtain $|\mathbf{x}| \leq \sqrt{d} \, \|\mathbf{x}\|_{\mathbf{P}^\circ}$, which means $\mathbf{P}^\circ \sqrt{d} \subset \mathbf{B}_2^d$. By duality we have $\mathbf{B}_2^d \subset \sqrt{d} \, \mathbf{P}$. Therefore, $d(\mathbf{K}, \mathbf{P}) \leq d(\mathbf{K}, \mathbf{L}) \, d(\mathbf{L}, \mathbf{P}) \leq (1 + \epsilon)\sqrt{d}$, and, hence, we have the following immediate consequence of Theorem 9.4.6.

Corollary 9.4.7 *For every* **o***-symmetric convex body* $\mathbf{K}$ *in* $\mathbb{R}^d$ *and every* $\epsilon \in (0, 1]$ *there exists an* **o***-symmetric convex polytope* $\mathbf{P}$ *in* $\mathbb{R}^d$ *with* M *vertices such that* $d(\mathbf{K}, \mathbf{P}) \leq (1 + \epsilon)\sqrt{d}$ *and*

$$M \leq C \, \epsilon^{-2} \, d \, \ln(2d),$$

where C *is an absolute constant.*

Corollary 9.4.7 implies the general upper estimate for $\operatorname{vein}(\mathbf{K})$.

Theorem 9.4.8 *For every centrally symmetric convex body* $\mathbf{K}$ *in* $\mathbb{R}^d$ *one has*

$$\operatorname{vein}(\mathbf{K}) \leq C \, d^{3/2} \, \ln(2d),$$

where C *is an absolute constant.*

Proof: Let $\mathbf{P}$ be a polytope given by Corollary 9.4.7 applied to $\mathbf{K}$ with $\epsilon = 1$. Then $d(\mathbf{K}, \mathbf{P}) \leq 2\sqrt{d}$. Clearly, $\operatorname{vein}(\mathbf{P}) \leq M$ (just take the $\mathbf{p}_i$s in the definition of $\operatorname{vein}(\cdot)$ to be the vertices of $\mathbf{P}$). Thus, by Lemma 9.4.1 we obtain $\operatorname{vein}(\mathbf{K}) \leq 2M\sqrt{d}$, which completes the proof.   □

10

Selected Proofs on Coverings by Planks and Cylinders

10.1 Proof of Theorem 4.1.7

10.1.1 On coverings of convex bodies by two planks

Lemma 10.1.1 *If a convex body* $\mathbf{K}$ *in* $\mathbb{E}^d, d \geq 2$ *is covered by the planks* $\mathbf{P}_1$ *and* $\mathbf{P}_2$*, then* $w_{\mathbf{C}}(\mathbf{P}_1) + w_{\mathbf{C}}(\mathbf{P}_2) \geq w_{\mathbf{C}}(\mathbf{K})$ *for any convex body* $\mathbf{C}$ *in* $\mathbb{E}^d$*.*

Proof: Let H_1 (resp., H_2) be one of the two hyperplanes which bound the plank $\mathbf{P}_1$ (resp., $\mathbf{P}_2$). If H_1 and H_2 are translates of each other, then the claim is obviously true. Thus, without loss of generality we may assume that $L := H_1 \cap H_2$ is a $(d-2)$-dimensional affine subspace of $\mathbb{E}^d$. Let $\mathbb{E}^2$ be the 2-dimensional linear subspace of $\mathbb{E}^d$ that is orthogonal to L. If $(\cdot)'$ denotes the (orthogonal) projection of $\mathbb{E}^d$ parallel to L onto $\mathbb{E}^2$, then obviously, $w_{\mathbf{C}'}(\mathbf{P}_1') = w_{\mathbf{C}}(\mathbf{P}_1)$, $w_{\mathbf{C}'}(\mathbf{P}_2') = w_{\mathbf{C}}(\mathbf{P}_2)$ and $w_{\mathbf{C}'}(\mathbf{K}') \geq w_{\mathbf{C}}(\mathbf{K})$. Thus, it is sufficient to prove that

$$w_{\mathbf{C}'}(\mathbf{P}_1') + w_{\mathbf{C}'}(\mathbf{P}_2') \geq w_{\mathbf{C}'}(\mathbf{K}').$$

In other words, it is sufficient to prove Lemma 10.1.1 for $d = 2$. Hence, in the rest of the proof, $\mathbf{K}, \mathbf{C}, \mathbf{P}_1, \mathbf{P}_2, H_1$, and H_2 mean the sets introduced and defined above, however, for $d = 2$. Now, we can make the following easy observation

$$w_{\mathbf{C}}(\mathbf{P}_1) + w_{\mathbf{C}}(\mathbf{P}_2) = \frac{w(\mathbf{P}_1)}{w(\mathbf{C}, H_1)} + \frac{w(\mathbf{P}_2)}{w(\mathbf{C}, H_2)}$$

$$= \frac{w(\mathbf{P}_1)}{w(\mathbf{K}, H_1)} \frac{w(\mathbf{K}, H_1)}{w(\mathbf{C}, H_1)} + \frac{w(\mathbf{P}_2)}{w(\mathbf{K}, H_2)} \frac{w(\mathbf{K}, H_2)}{w(\mathbf{C}, H_2)}$$

$$\geq \left(\frac{w(\mathbf{P}_1)}{w(\mathbf{K}, H_1)} + \frac{w(\mathbf{P}_2)}{w(\mathbf{K}, H_2)} \right) w_{\mathbf{C}}(\mathbf{K})$$

$$= (w_{\mathbf{K}}(\mathbf{P}_1) + w_{\mathbf{K}}(\mathbf{P}_2)) w_{\mathbf{C}}(\mathbf{K}).$$

K. Bezdek, *Classical Topics in Discrete Geometry*, CMS Books in Mathematics, DOI 10.1007/978-1-4419-0600-7_10, © Springer Science+Business Media, LLC 2010

Then recall that Theorem 4 in [3] states that if a convex set in the plane is covered by planks, then the sum of their relative widths is at least 1. Thus, using our terminology, we have that $w_{\mathbf{K}}(\mathbf{P}_1) + w_{\mathbf{K}}(\mathbf{P}_2) \geq 1$, finishing the proof of Lemma 10.1.1.   □

10.1.2 A proof of the affine plank conjecture of Bang for non-overlapping cuts

Let $\mathbf{K}$ and $\mathbf{C}$ be convex bodies in $\mathbb{E}^d$, $d \geq 2$. We prove Theorem 4.1.7 by induction on n. It is trivial to check the claim for $n = 1$. So, let $n \geq 2$ be given and assume that Theorem 4.1.7 holds for at most $n - 2$ successive hyperplane cuts and based on that we show that it holds for $n - 1$ successive hyperplane cuts as well. The details are as follows.

Let $H_1, \ldots, H_{n-1}$ denote the hyperplanes of the $n - 1$ successive hyperplane cuts that slice $\mathbf{K}$ into n pieces such that the greatest $\mathbf{C}$ inradius of the pieces is the smallest possible say, ρ. Then take the first cut H_1 that slices $\mathbf{K}$ into the pieces $\mathbf{K}_1$ and $\mathbf{K}_2$ such that $\mathbf{K}_1$ (resp., $\mathbf{K}_2$) is sliced into n_1 (resp., n_2) pieces by the successive hyperplane cuts $H_2, \ldots, H_{n-1}$, where $n = n_1 + n_2$. The induction hypothesis implies that $\rho \geq r_{\mathbf{C}}(\mathbf{K}_1, n_1) =: \rho_1$ and $\rho \geq r_{\mathbf{C}}(\mathbf{K}_2, n_2) =: \rho_2$ and therefore

$$w_{\mathbf{C}}(\mathbf{K}_1{}^{\rho\mathbf{C}}) \leq w_{\mathbf{C}}(\mathbf{K}_1{}^{\rho_1\mathbf{C}}) = n_1\rho_1 \leq n_1\rho; \tag{10.1}$$

moreover,

$$w_{\mathbf{C}}(\mathbf{K}_2{}^{\rho\mathbf{C}}) \leq w_{\mathbf{C}}(\mathbf{K}_2{}^{\rho_2\mathbf{C}}) = n_2\rho_2 \leq n_2\rho. \tag{10.2}$$

Now, we need to define the following set. In order to simplify matters let us assume that the origin $\mathbf{o}$ of $\mathbb{E}^d$ belongs to the interior of $\mathbf{C}$. Then consider all translates of $\rho\mathbf{C}$ which are contained in $\mathbf{K}$. The set of points in the translates of $\rho\mathbf{C}$ that correspond to $\mathbf{o}$ form a convex set called the *inner $\rho\mathbf{C}$-parallel body of* $\mathbf{K}$ denoted by $\mathbf{K}_{-\rho\mathbf{C}}$.

Clearly,

$$(\mathbf{K}_1)_{-\rho\mathbf{C}} \cup (\mathbf{K}_2)_{-\rho\mathbf{C}} \subset \mathbf{K}_{-\rho\mathbf{C}} \text{ with } (\mathbf{K}_1)_{-\rho\mathbf{C}} \cap (\mathbf{K}_2)_{-\rho\mathbf{C}} = \emptyset.$$

Also, it is easy to see that there is a plank $\mathbf{P}$ with $w_{\mathbf{C}}(\mathbf{P}) = \rho$ such that it is parallel to H_1 and contains H_1 in its interior; moreover,

$$\mathbf{K}_{-\rho\mathbf{C}} \subset (\mathbf{K}_1)_{-\rho\mathbf{C}} \cup (\mathbf{K}_2)_{-\rho\mathbf{C}} \cup \mathbf{P}.$$

Hence, applying Lemma 10.1.1 to $(\mathbf{K}_1)_{-\rho\mathbf{C}}$ as well as $(\mathbf{K}_2)_{-\rho\mathbf{C}}$ we get that

$$w_{\mathbf{C}}(\mathbf{K}_{-\rho\mathbf{C}}) \leq w_{\mathbf{C}}((\mathbf{K}_1)_{-\rho\mathbf{C}}) + \rho + w_{\mathbf{C}}((\mathbf{K}_2)_{-\rho\mathbf{C}}). \tag{10.3}$$

It follows from the definitions that $w_{\mathbf{C}}((\mathbf{K}_1)_{-\rho\mathbf{C}}) = w_{\mathbf{C}}(\mathbf{K}_1{}^{\rho\mathbf{C}}) - \rho$, $w_{\mathbf{C}}((\mathbf{K}_2)_{-\rho\mathbf{C}}) = w_{\mathbf{C}}(\mathbf{K}_2{}^{\rho\mathbf{C}}) - \rho$ and $w_{\mathbf{C}}(\mathbf{K}_{-\rho\mathbf{C}}) = w_{\mathbf{C}}(\mathbf{K}^{\rho\mathbf{C}}) - \rho$. Hence, (10.3) is equivalent to

$$w_{\mathbf{C}}(\mathbf{K}^{\rho\mathbf{C}}) \leq w_{\mathbf{C}}(\mathbf{K}_1{}^{\rho\mathbf{C}}) + w_{\mathbf{C}}(\mathbf{K}_2{}^{\rho\mathbf{C}}). \tag{10.4}$$

Finally, (10.1),(10.2), and (10.4) yield that

$$w_{\mathbf{C}}(\mathbf{K}^{\rho\mathbf{C}}) \leq n_1\rho + n_2\rho = n\rho. \tag{10.5}$$

Thus, (10.4) clearly implies that $r_{\mathbf{C}}(\mathbf{K}, n) \leq \rho$. As the case, when the optimal partition is achieved, follows directly from the definition of the nth successive $\mathbf{C}$-inradius of $\mathbf{K}$, the proof of Theorem 4.1.7 is complete.

10.2 Proof of Theorem 4.2.2

10.2.1 Covering ellipsoids by 1-codimensional cylinders

We prove the part of Theorem 4.2.2 on ellipsoids. In this case, every $\mathbf{C}_i$ can be presented as $\mathbf{C}_i = l_i + B_i$, where l_i is a line containing the origin $\mathbf{o}$ in $\mathbb{E}^d$ and B_i is a measurable set in $E_i := l_i^\perp$. Because $\mathrm{crv}_{\mathbf{K}}(\mathbf{C}) = \mathrm{crv}_{T\mathbf{K}}(T\mathbf{C})$ for every invertible affine map $T : \mathbb{E}^d \to \mathbb{E}^d$, we therefore may assume that $\mathbf{K} = \mathbf{B}^d$, where $\mathbf{B}^d$ denotes the unit ball centered at the origin $\mathbf{o}$ in $\mathbb{E}^d$. Recall that $\omega_{d-1} := \mathrm{vol}_{d-1}(\mathbf{B}^{d-1})$. Then

$$\mathrm{crv}_{\mathbf{K}}(\mathbf{C}_i) = \frac{\mathrm{vol}_{d-1}(B_i)}{\omega_{d-1}}.$$

Consider the following (density) function on $\mathbb{E}^d$,

$$p(\mathbf{x}) = 1/\sqrt{1 - \|\mathbf{x}\|^2}$$

for $\|\mathbf{x}\| < 1$ and $p(\mathbf{x}) = 0$ otherwise, where $\|\ \|$ denotes the standard Euclidean norm in $\mathbb{E}^d$. The corresponding measure on $\mathbb{E}^d$ we denote by μ; that is, $d\mu(\mathbf{x}) = p(\mathbf{x})d\mathbf{x}$. Let l be a line containing $\mathbf{o}$ in $\mathbb{E}^d$ and $E = l^\perp$. It follows from direct calculations that for every $\mathbf{z} \in E$ with $\|\mathbf{z}\| < 1$,

$$\int_{l+\mathbf{z}} p(\mathbf{x})\ d\mathbf{x} = \pi.$$

Thus, we have

$$\mu(\mathbf{B}^d) = \int_{\mathbf{B}^d} p(\mathbf{x})\ d\mathbf{x} = \int_{\mathbf{B}^d \cap E} \int_{l+\mathbf{z}} p(\mathbf{x})\ d\mathbf{x}\ d\mathbf{z} = \pi\,\omega_{d-1}$$

and for every $i \leq N$,

$$\mu(\mathbf{C}_i) = \int_{\mathbf{C}_i} p(\mathbf{x})\ d\mathbf{x} = \int_{B_i} \int_{l_i+\mathbf{z}} p(\mathbf{x})\ d\mathbf{x}\ d\mathbf{z} = \pi\,\mathrm{vol}_{d-1}(B_i).$$

Because $\mathbf{B}^d \subset \bigcup_{i=1}^{N} \mathbf{C}_i$, we obtain

$$\pi\,\omega_{d-1} = \mu(\mathbf{B}^d) \le \mu\left(\bigcup_{i=1}^{N}\mathbf{C}_i\right) \le \sum_{i=1}^{N}\mu(\mathbf{C}_i) = \sum_{i=1}^{N}\pi\,\mathrm{vol}_{d-1}(B_i).$$

It implies

$$\sum_{i=1}^{N}\mathrm{crv}_{\mathbf{B}^d}(\mathbf{C}_i) = \sum_{i=1}^{N}\frac{\mathrm{vol}_{d-1}(B_i)}{\omega_{d-1}} \ge 1.$$

10.2.2 Covering convex bodies by cylinders of given codimension

We show the general case of Theorem 4.2.2 as follows. For $i \le N$ denote $\bar{\mathbf{C}}_i := \mathbf{C}_i \cap \mathbf{K}$ and $E_i := H_i^{\perp}$ and note that

$$\mathbf{K} \subset \bigcup_{i=1}^{N}\bar{\mathbf{C}}_i \qquad \text{and} \qquad P_{E_i}\bar{\mathbf{C}}_i = B_i \cap P_{E_i}\mathbf{K}.$$

Because $\bar{\mathbf{C}}_i \subset \mathbf{K}$ we also have

$$\max_{\mathbf{x}\in\mathbb{E}^d}\ \mathrm{vol}_k(\bar{\mathbf{C}}_i \cap (\mathbf{x}+H_i)) \le \max_{\mathbf{x}\in\mathbb{E}^d}\ \mathrm{vol}_k(\mathbf{K}\cap(\mathbf{x}+H_i)).$$

We use the following theorem, proved by Rogers and Shephard [221] (see also [102] and Lemma 8.8 in [215]).

Theorem 10.2.1 *Let* $1 \le k \le d-1$. *Let* $\mathbf{K}$ *be a convex body in* $\mathbb{E}^d$ *and* E *be a* k-*dimensional linear subspace of* $\mathbb{E}^d$. *Then*

$$\max_{\mathbf{x}\in\mathbb{E}^d}\ \mathrm{vol}_{d-k}(\mathbf{K}\cap(\mathbf{x}+E^{\perp}))\,\mathrm{vol}_k(P_E\mathbf{K}) \le \binom{d}{k}\mathrm{vol}_d(\mathbf{K}).$$

We note that the reverse estimate

$$\max_{\mathbf{x}\in\mathbb{E}^d}\ \mathrm{vol}_{d-k}(\mathbf{K}\cap(\mathbf{x}+E^{\perp}))\,\mathrm{vol}_k(P_E\mathbf{K}) \ge \mathrm{vol}_d(\mathbf{K})$$

is a simple application of the Fubini theorem and is correct for any measurable set $\mathbf{K}$ in $\mathbb{E}^d$.

Thus, applying Theorem 10.2.1 (and the remark after it, saying that we don't need convexity of $\bar{\mathbf{C}}_i$) we obtain for every $i \le N$:

$$\mathrm{crv}_{\mathbf{K}}(\mathbf{C}_i) = \frac{\mathrm{vol}_{d-k}(B_i)}{\mathrm{vol}_{d-k}(P_{E_i}\mathbf{K})} \ge \frac{\mathrm{vol}_{d-k}(P_{E_i}\bar{\mathbf{C}}_i)}{\mathrm{vol}_{d-k}(P_{E_i}\mathbf{K})}$$

$$\ge \frac{\mathrm{vol}_d(\bar{\mathbf{C}}_i)}{\max_{\mathbf{x}\in\mathbb{E}^d}\ \mathrm{vol}_k(\bar{\mathbf{C}}_i\cap(\mathbf{x}+H_i))} \frac{\max_{\mathbf{x}\in\mathbb{E}^d}\ \mathrm{vol}_k(\mathbf{K}\cap(\mathbf{x}+H_i))}{\binom{d}{k}\mathrm{vol}_d(\mathbf{K})}$$

$$\ge \frac{\mathrm{vol}_d(\bar{\mathbf{C}}_i)}{\binom{d}{k}\mathrm{vol}_d(\mathbf{K})}.$$

Using that $\bar{\mathbf{C}}_i$s cover $\mathbf{K}$, we observe

$$\sum_{i=1}^{N} \operatorname{crv}_{\mathbf{K}}(\mathbf{C}_i) \geq \frac{1}{\binom{d}{k}},$$

which completes the proof.

10.3 Proof of Theorem 4.5.2

Let $\mathbf{P}_1, \mathbf{P}_2, \ldots, \mathbf{P}_n$ be an arbitrary family of planks of width $w_1, w_2, \ldots, w_n$ in $\mathbb{E}^3$ and let $\mathbf{P}$ be a plank of width $w_1 + w_2 + \cdots + w_n$ with $\mathbf{o}$ as a center of symmetry. Moreover, let $S(x)$ denote the sphere of radius x centered at $\mathbf{o}$. Now, recall the well-known fact that if $\mathbf{P}(y)$ is a plank of width y whose both boundary planes intersect $S(x)$, then $\operatorname{sarea}(S(x) \cap \mathbf{P}(y)) = 2\pi xy$, where $\operatorname{sarea}(\ . \ )$ refers to the surface area measure on $S(x)$. This implies in a straightforward way that

$$\operatorname{sarea}[(\mathbf{P}_1 \cup \mathbf{P}_2 \cup \cdots \cup \mathbf{P}_n) \cap S(x)] \leq \operatorname{sarea}(\mathbf{P} \cap S(x)),$$

and so,

$$\operatorname{vol}_3((\mathbf{P}_1 \cup \mathbf{P}_2 \cup \cdots \cup \mathbf{P}_n) \cap \mathbf{B}^3) = \int_0^1 \operatorname{sarea}[(\mathbf{P}_1 \cup \mathbf{P}_2 \cup \cdots \cup \mathbf{P}_n) \cap S(x)] \ dx$$

$$\leq \int_0^1 \operatorname{sarea}(\mathbf{P} \cap S(x)) \ dx = \operatorname{vol}_3(\mathbf{P} \cap \mathbf{B}^3),$$

finishing the proof of the "if" part of Theorem 4.5.2. Actually, a closer look of the above argument gives a proof of the "only if" part as well.

10.4 Proof of Theorem 4.5.8

Recall that if X is a finite set lying in the interior of a unit ball in $\mathbb{E}^d$, then we can talk about its spindle convex hull $\operatorname{conv}_s(X)$, which is simply the intersection of all (closed) unit balls of $\mathbb{E}^d$ that contain X (for more details see [69]). The following statement can be obtained by combining Corollary 3.4 of [69] and Proposition 1 of [65].

Lemma 10.4.1 *Let X be a finite set lying in the interior of a unit ball in $\mathbb{E}^d$. Then*
(i) $\operatorname{conv}_s(X) = \mathbf{B}[\mathbf{B}[X]]$ and therefore $\mathbf{B}[X] = \mathbf{B}[\operatorname{conv}_s(X)]$,
(ii) the Minkowski sum $\mathbf{B}[X] + \operatorname{conv}_s(X)$ is a convex body of constant width 2 in $\mathbb{E}^d$ and so, $w(\mathbf{B}[X]) + \operatorname{diam}(\operatorname{conv}_s(X)) = 2$, where $\operatorname{diam}(\ . \ )$ stands for the diameter of the corresponding set in $\mathbb{E}^d$.

By part (ii) of Lemma 10.4.1, $\mathrm{diam}\big(\mathrm{conv}_s(X)\big) \leq 2-x$. This implies, via a classical theorem of convexity (see, e.g., [85]), the existence of a convex body $\mathbf{L}$ of constant width $(2-x)$ in $\mathbb{E}^d$ with $\mathrm{conv}_s(X) \subset \mathbf{L}$. Hence, using part (i) of Lemma 10.4.1, we get that $\mathbf{B}[\mathbf{L}] \subset \mathbf{B}[X] = \mathbf{B}\big[\mathrm{conv}_s(X)\big]$. Finally, notice that as $\mathbf{L}$ is a convex body of constant width $(2-x)$ therefore $\mathbf{B}[\mathbf{L}]$ is, in fact, the outer-parallel domain of $\mathbf{L}$ having radius $(x-1)$ (i.e., $\mathbf{B}[\mathbf{L}]$ is the union of all d-dimensional (closed) balls of radii $(x-1)$ in $\mathbb{E}^d$ that are centered at the points of $\mathbf{L}$). Thus,

$$\mathrm{vol}_d(\mathbf{B}[X]) \geq \mathrm{vol}_d(\mathbf{B}[\mathbf{L}]) = \mathrm{vol}_d(\mathbf{L}) + \mathrm{svol}_{d-1}(\mathrm{bd}(\mathbf{L}))(x-1) + \mathrm{vol}_d(\mathbf{B}^d)(x-1)^d.$$

The above inequality together with the following obvious ones

$$\mathrm{vol}_d(\mathbf{L}) \geq \mathrm{vol}_d(\mathbf{K}_{BL}^{2-x,d}) \text{ and } \mathrm{svol}_{d-1}(\mathrm{bd}(\mathbf{L})) \geq \mathrm{svol}_{d-1}(\mathrm{bd}(\overline{\mathbf{K}}_{BL}^{2-x,d}))$$

implies Theorem 4.5.8 in a straightforward way.

Selected Proofs on the Kneser–Poulsen Conjecture

11.1 Proof of Theorem 5.3.2 on the Monotonicity of Weighted Surface Volume

First, recall the following underlying system of (truncated) Voronoi cells. For a given point configuration $\mathbf{p} = (\mathbf{p}_1, \mathbf{p}_2, \ldots, \mathbf{p}_N)$ in $\mathbb{E}^d$ and radii $r_1, r_2, \ldots,$ r_N consider the following sets,

$$\mathbf{V}_i = \{\mathbf{x} \in \mathbb{E}^d \mid \text{for all } j, \ \|\mathbf{x} - \mathbf{p}_i\|^2 - r_i^2 \leq \|\mathbf{x} - \mathbf{p}_j\|^2 - r_j^2\},$$

$$\mathbf{V}^i = \{\mathbf{x} \in \mathbb{E}^d \mid \text{for all } j, \ \|\mathbf{x} - \mathbf{p}_i\|^2 - r_i^2 \geq \|\mathbf{x} - \mathbf{p}_j\|^2 - r_j^2\}.$$

The set $\mathbf{V}_i$ (resp., $\mathbf{V}^i$) is called the *nearest (resp., farthest) point Voronoi cell* of the point $\mathbf{p}_i$. We now restrict each of these sets as follows,

$$\mathbf{V}_i(r_i) = \mathbf{V}_i \cap \mathbf{B}^d[\mathbf{p}_i, r_i],$$

$$\mathbf{V}^i(r_i) = \mathbf{V}^i \cap \mathbf{B}^d[\mathbf{p}_i, r_i].$$

We call the set $\mathbf{V}_i(r_i)$ (resp., $\mathbf{V}^i(r_i)$) the *nearest (resp., farthest) point truncated Voronoi cell* of the point $\mathbf{p}_i$. For each $i \neq j$ let $W_{ij} = \mathbf{V}_i \cap \mathbf{V}_j$ and $W^{ij} = \mathbf{V}^i \cap \mathbf{V}^j$. The sets W_{ij} and W^{ij} are the *walls* between the nearest point and farthest point Voronoi cells. Finally, it is natural to define the relevant *truncated walls* as follows.

$$W_{ij}(\mathbf{p}_i, r_i) = W_{ij} \cap \mathbf{B}^d[\mathbf{p}_i, r_i]$$
$$= W_{ij}(\mathbf{p}_j, r_j) = W_{ij} \cap \mathbf{B}^d[\mathbf{p}_j, r_j],$$

$$W^{ij}(\mathbf{p}_i, r_i) = W^{ij} \cap \mathbf{B}^d[\mathbf{p}_i, r_i]$$
$$= W^{ij}(\mathbf{p}_j, r_j) = W^{ij} \cap \mathbf{B}^d[\mathbf{p}_j, r_j].$$

K. Bezdek, *Classical Topics in Discrete Geometry*, CMS Books in Mathematics,
DOI 10.1007/978-1-4419-0600-7_11, © Springer Science+Business Media, LLC 2010

Second, for each $i = 1, 2, \ldots, N$ and $0 \le s$, define $r_i(s) = \sqrt{r_i^2 + s}$. Clearly,

$$\frac{d}{ds} r_i(s) = \frac{1}{2r_i(s)}. \tag{11.1}$$

Now, define $\mathbf{r}(s) = (r_1(s), \ldots, r_N(s))$, and introduce

$$V_d(t, s) := \mathrm{vol}_d \left(\mathbf{B}_{\cup}^d[\mathbf{p}(t), \mathbf{r}(s)] \right),$$

and

$$V^d(t, s) := \mathrm{vol}_d \left(\mathbf{B}_{\cap}^d[\mathbf{p}(t), \mathbf{r}(s)] \right)$$

as functions of the variables t and s, where

$$\mathbf{B}_{\cup}^d[\mathbf{p}(t), \mathbf{r}(s)] := \bigcup_{i=1}^{N} \mathbf{B}^d[\mathbf{p}_i(t), r_i(s)],$$

and

$$\mathbf{B}_{\cap}^d[\mathbf{p}(t), \mathbf{r}(s)] := \bigcap_{i=1}^{N} \mathbf{B}^d[\mathbf{p}_i(t), r_i(s)].$$

Throughout we assume that all $r_i > 0$.

Lemma 11.1.1 *Let $d \ge 2$ and let $\mathbf{p}(t), 0 \le t \le 1$ be a smooth motion of a point configuration in $\mathbb{E}^d$ such that for each t, the points of the configuration are pairwise distinct. Then the volume functions $V_d(t, s)$ and $V^d(t, s)$ are continuously differentiable in t and s simultaneously, and for any fixed t, the nearest point and farthest point Voronoi cells are constant.*

Proof: Let $t = t_0$ be fixed. Then recall that the point $\mathbf{x}$ belongs to the Voronoi cell $\mathbf{V}_i(t_0, s)$ (resp., $\mathbf{V}^i(t_0, s)$), when for all j, $\|\mathbf{x} - \mathbf{p}_i(t_0)\|^2 - \|\mathbf{x} - \mathbf{p}_j(t_0)\|^2 - r_i(s)^2 + r_j(s)^2$ is non-positive (resp., non-negative). But $r_i(s)^2 - r_j(s)^2 = r_i^2 - r_j^2$ is constant. So each $\mathbf{V}_i(t_0, s)$ and $\mathbf{V}^i(t_0, s)$ is a constant function of s.

As $\mathbf{p}(t)$ is continuously differentiable, therefore the partial derivatives of $V_d(t, s)$ and $V^d(t, s)$ with respect to t exist and are continuous by Theorem 5.2.1. Each ball $\mathbf{B}^d[\mathbf{p}_i(t), r_i(s)], d \ge 2$ is strictly convex. Hence, the $(d - 1)$-dimensional surface volume of the boundaries of $\mathbf{B}_{\cup}^d[\mathbf{p}(t), \mathbf{r}(s)]$ and $\mathbf{B}_{\cap}^d[\mathbf{p}(t), \mathbf{r}(s)]$ are continuous functions of s, and the partial derivatives of $V_d(t, s)$ and $V^d(t, s)$ with respect to s exist and are continuous. Thus, $V_d(t, s)$ and $V^d(t, s)$ are both continuously differentiable with respect to t and s simultaneously. $\qquad\square$

Lemma 11.1.2 *Let $\mathbf{p}(t), 0 \le t \le 1$ be an analytic motion of a point configuration in $\mathbb{E}^d, d \ge 2$. Then there exists an open dense set U in $[0, 1] \times (0, \infty)$ such that for any $(t, s) \in U$ the following hold.*

$$\frac{\partial^2}{\partial t \partial s} V_d(t,s) = \sum_{1 \leq i < j \leq N} \left(\frac{d}{dt} d_{ij}(t) \right) \cdot \frac{\partial}{\partial s} \text{vol}_{d-1} \left[W_{ij}(\mathbf{p}_i(t), r_i(s)) \right],$$

and

$$\frac{\partial^2}{\partial t \partial s} V^d(t,s) = \sum_{1 \leq i < j \leq N} - \left(\frac{d}{dt} d_{ij}(t) \right) \cdot \frac{\partial}{\partial s} \text{vol}_{d-1} \left[W^{ij}(\mathbf{p}_i(t), r_i(s)) \right].$$

Hence, if $\mathbf{p}(t)$ *is contracting, then* $\frac{\partial}{\partial s} V_d(t,s)$ *is monotone decreasing in* t, *and* $\frac{\partial}{\partial s} V^d(t,s)$ *is monotone increasing in* t.

Proof: Given that $\mathbf{p}(t), 0 \leq t \leq 1$ is an analytic function of t, we wish to define an open dense set U in $[0,1] \times (0, \infty)$, where the volume functions $V_d(t,s)$ and $V^d(t,s)$ are analytic in t and s simultaneously. Lemma 11.1.1 implies that the Voronoi cells $\mathbf{V}_i$ and $\mathbf{V}^i$ are functions of t alone. Moreover, clearly there are only a finite number of values of t in the interval $[0,1]$, where the combinatorial type of the above Voronoi cells changes. The volume of the truncated Voronoi cells $\mathbf{V}_i (r_i(s))$ and $\mathbf{V}^i (r_i(s))$ are obtained from the volume of the d-dimensional Euclidean ball of radius $r_i(s)$ by removing or adding the volumes of the regions obtained by conning over the walls $W_{ij}(\mathbf{p}_i(t), r_i(s))$ or $W^{ij}(\mathbf{p}_i(t), r_i(s))$ from the point $\mathbf{p}_i(t)$. By induction on d, starting at $d = 1$, each W_{ij} and W^{ij} is an analytic function of t and s, when the ball of radius $r_i(s)$ is not tangent to any of the faces of $\mathbf{V}_i$ or $\mathbf{V}^i$. So, for any fixed t the ball of radius $r_i(s)$ will not be tangent to any of the faces $\mathbf{V}_i$ or $\mathbf{V}^i$ for all but a finite number of values of s. Thus, we define U to be the set of those (t,s), where for some open interval about t in $[0,1]$, the combinatorial type of the Voronoi cells is constant and for all i, the ball of radius $r_i(s)$ is not tangent to any of the faces of $\mathbf{V}_i$ or $\mathbf{V}^i$. We also assume that the points of the configuration $\mathbf{p}(t)$ are distinct for any $(t,s) \in U$. If, for $i \neq j$ and for infinitely many values of t in the interval $[0,1]$, $\mathbf{p}_i(t) = \mathbf{p}_j(t)$, then they are the same point for all t, and those points may be identified. Then the set U is open and dense in $[0,1] \times (0, \infty)$ and the volume functions $V_d(t,s)$ and $V^d(t,s)$ are analytic in t and s. Thus, the formulas for the mixed partial derivatives in Lemma 11.1.2 follow from the definition of U and from Theorem 5.2.1. (Note also that here we could interchange the order of partial differentiation with respect to the variables t and s.)

To show that $\frac{\partial}{\partial s} V_d(t,s)$ and $\frac{\partial}{\partial s} V^d(t,s)$ are monotone, suppose they are not. We show a contradiction. If we perturb s slightly to s_0 say, then using the formulas for the mixed partial derivatives in Lemma 11.1.2 we get that the partial derivative of $\frac{\partial}{\partial s} V_d(t,s)$ and $\frac{\partial}{\partial s} V^d(t,s)$ with respect to t exists and has the appropriate sign, except for a finite number of values of t for $s = s_0$. (Here we have also used the following rather obvious monotonicity property of the walls: $W_{ij}(\mathbf{p}_i(t), r_i(s)) \subset W_{ij}(\mathbf{p}_i(t), r_i(s^*))$ and $W^{ij}(\mathbf{p}_i(t), r_i(s)) \subset W^{ij}(\mathbf{p}_i(t), r_i(s^*))$ for any $s \leq s^*$.) Since $\frac{\partial}{\partial s} V_d(t,s)$ and $\frac{\partial}{\partial s} V^d(t,s)$ are continuous as a function of t at $s = s_0$ by Lemma 11.1.1, they are monotone. But the functions at s_0 approximate the functions at s (again by Lemma 11.1.1) providing the contradiction. So, $\frac{\partial}{\partial s} V_d(t,s)$ and $\frac{\partial}{\partial s} V^d(t,s)$ are indeed monotone. This completes the proof of Lemma 11.1.2.   □

First, note that

$$F_i(t) = \mathbf{V}_i(t,0) \cap \mathrm{bd}\left(\mathbf{B}_\cup^d[\mathbf{p}(t), \mathbf{r}(0)]\right) \tag{11.2}$$

and

$$\left(\text{resp., } F_i(t) = \mathbf{V}^i(t,0) \cap \mathrm{bd}\left(\mathbf{B}_\cap^d[\mathbf{p}(t), \mathbf{r}(0)]\right)\right). \tag{11.3}$$

Second, (11.1), (11.2) and (11.3) imply in a straightforward way that

$$\frac{\partial}{\partial s} V_d(t,s)\Big|_{s=0} = \frac{1}{2} \sum_{i=1}^{N} \frac{1}{r_i} \mathrm{svol}_{d-1}(F_i(t)) = \lim_{s_0 \to 0^+} \frac{\partial}{\partial s} V_d(t,s)\Big|_{s=s_0} \tag{11.4}$$

$$\left(\text{resp., } \frac{\partial}{\partial s} V^d(t,s)\Big|_{s=0} = \frac{1}{2} \sum_{i=1}^{N} \frac{1}{r_i} \mathrm{svol}_{d-1}(F_i(t)) = \lim_{s_0 \to 0^+} \frac{\partial}{\partial s} V_d(t,s)\Big|_{s=s_0}\right). \tag{11.5}$$

Thus, (11.4) and (11.5) together with Lemma 11.1.2 finish the proof of Theorem 5.3.2.

11.2 Proof of Theorem 5.3.3 on Weighted Surface and Codimension Two Volumes

We start with the following volume formula from calculus, which is based on cylindrical shells.

Lemma 11.2.1 *Let X be a compact measurable set in $\mathbb{E}^d, d \geq 3$ that is a solid of revolution about $\mathbb{E}^{d-2}$. In other words the orthogonal projection of $X \cap \{\mathbb{E}^{d-2} \times (s\cos\theta, s\sin\theta)\}$ onto $\mathbb{E}^{d-2}$ is a measurable set $X(s)$ independent of θ. Then*

$$\mathrm{vol}_d(X) = \int_0^\infty (2\pi s)\mathrm{vol}_{d-2}(X(s))\,ds.$$

By assumption the centers of the closed d-dimensional balls $\mathbf{B}^d[\mathbf{p}_i, r_i]$, $1 \leq i \leq N$ lie in the $(d-2)$-dimensional affine subspace L of $\mathbb{E}^d$. Now, recall the construction of the following (truncated) Voronoi cells.

$$\mathbf{V}_i(d) = \{\mathbf{x} \in \mathbb{E}^d \mid \text{for all } j, \ \|\mathbf{x} - \mathbf{p}_i\|^2 - r_i^2 \leq \|\mathbf{x} - \mathbf{p}_j\|^2 - r_j^2\},$$

$$\mathbf{V}^i(d) = \{\mathbf{x} \in \mathbb{E}^d \mid \text{for all } j, \ \|\mathbf{x} - \mathbf{p}_i\|^2 - r_i^2 \geq \|\mathbf{x} - \mathbf{p}_j\|^2 - r_j^2\}.$$

The set $\mathbf{V}_i(d)$ (resp., $\mathbf{V}^i(d)$) is called the nearest (resp., farthest) point Voronoi cell of the point $\mathbf{p}_i$ in $\mathbb{E}^d$. Then we restrict each of these sets as follows:

$$\mathbf{V}_i(r_i, d) = \mathbf{V}_i \cap \mathbf{B}^d[\mathbf{p}_i, r_i],$$

$$\mathbf{V}^i(r_i, d) = \mathbf{V}^i \cap \mathbf{B}^d[\mathbf{p}_i, r_i].$$

We call the set $\mathbf{V}_i(r_i, d)$ (resp., $\mathbf{V}^i(r_i, d)$) the nearest (resp., farthest) point truncated Voronoi cell of the point $\mathbf{p}_i$ in $\mathbb{E}^d$. As the point configuration $\mathbf{p} = (\mathbf{p}_1, \mathbf{p}_2, \ldots, \mathbf{p}_N)$ lies in the $(d-2)$-dimensional affine subspace $L \subset \mathbb{E}^d$ and as without loss of generality we may assume that $L = \mathbb{E}^{d-2}$, therefore one can introduce the relevant $(d-2)$-dimensional truncated Voronoi cells $\mathbf{V}_i(r_i, d-2)$ and $\mathbf{V}^i(r_i, d-2)$ in a straightforward way. We are especially interested in the relation of the volume of $\mathbf{V}_i(r_i, d-2)$ and $\mathbf{V}^i(r_i, d-2)$ in $\mathbb{E}^{d-2}$ to the volume of the corresponding truncated Voronoi cells $\mathbf{V}_i(r_i, d)$ and $\mathbf{V}^i(r_i, d)$ in $\mathbb{E}^d$.

Lemma 11.2.2 *We have that*

$$\mathrm{vol}_d \left(\mathbf{V}_i(r_i, d) \right) = \int_0^{r_i} (2\pi s) \mathrm{vol}_{d-2} \left(\mathbf{V}_i(s, d-2) \right) ds,$$

and

$$\mathrm{vol}_d \left(\mathbf{V}^i(r_i, d) \right) = \int_0^{r_i} (2\pi s) \mathrm{vol}_{d-2} \left(\mathbf{V}^i(s, d-2) \right) ds.$$

Proof: It is clear, in both cases, that $\mathbf{V}_i(r_i, d)$ and $\mathbf{V}^i(r_i, d)$ are compact measurable sets of revolution (about $\mathbb{E}^{d-2}$). Note that the orthogonal projection of $\mathbf{B}^d[\mathbf{p}_i, r_i] \cap \{\mathbb{E}^{d-2} \times (s\cos\theta, s\sin\theta)\}$ onto $\mathbb{E}^{d-2}$ is the $(d-2)$-dimensional ball of radius $\sqrt{r_i^2 - s^2}$ centered at $\mathbf{p}_i$. Thus, by Lemma 11.2.1 we have that

$$\mathrm{vol}_d \left(\mathbf{V}_i(r_i, d) \right) = \int_0^{r_i} (2\pi s) \mathrm{vol}_{d-2} \left(\mathbf{V}_i \left(\sqrt{r_i^2 - s^2}, d-2 \right) \right) ds \ .$$

But if we make the change of variable $u = \sqrt{r_i^2 - s^2}$, we get the desired integral. A similar calculation works for $\mathrm{vol}_d \left(\mathbf{V}^i(r_i, d) \right)$. $\quad\square$

The following is an immediate corollary of Lemma 11.2.2.

Corollary 11.2.3 *We have that*

$$\frac{d}{dr} \mathrm{vol}_d \left(\mathbf{V}_i(r, d) \right) \bigg|_{r=r_i} = 2\pi r_i \mathrm{vol}_{d-2} \left(\mathbf{V}_i(r_i, d-2) \right),$$

and

$$\frac{d}{dr} \mathrm{vol}_d \left(\mathbf{V}^i(r, d) \right) \bigg|_{r=r_i} = 2\pi r_i \mathrm{vol}_{d-2} \left(\mathbf{V}^i(r_i, d-2) \right).$$

Moreover, it is clear that if F_i stands for the contribution of the ith ball to the boundary of the union $\bigcup_{i=1}^{N} \mathbf{B}^d[\mathbf{p}_i, r_i]$, then

$$\mathrm{svol}_{d-1}(F_i) = \frac{d}{dr} \mathrm{vol}_d \left(\mathbf{V}_i(r, d) \right) \Big|_{r=r_i}. \tag{11.6}$$

Similarly, if F_i denotes the contribution of the ith ball to the boundary of the intersection $\bigcap_{i=1}^{N} \mathbf{B}^d[\mathbf{p}_i, r_i]$, then

$$\mathrm{svol}_{d-1}(F_i) = \frac{d}{dr} \mathrm{vol}_d \left(\mathbf{V}^i(r, d) \right) \Big|_{r=r_i}. \tag{11.7}$$

Finally, it is obvious that

$$\mathrm{vol}_{d-2} \left(\bigcup_{i=1}^{N} \mathbf{B}^{d-2}[\mathbf{p}_i, r_i] \right) = \sum_{i=1}^{N} \mathrm{vol}_{d-2} \left(\mathbf{V}_i(r_i, d-2) \right), \tag{11.8}$$

and

$$\mathrm{vol}_{d-2} \left(\bigcap_{i=1}^{N} \mathbf{B}^{d-2}[\mathbf{p}_i, r_i] \right) = \sum_{i=1}^{N} \mathrm{vol}_{d-2} \left(\mathbf{V}^i(r_i, d-2) \right). \tag{11.9}$$

Thus, Corollary 11.2.3 and (11.6), (11.8) (resp., (11.7), (11.9)) finish the proof of Theorem 5.3.3.

11.3 Proof of Theorem 5.3.4 - the Leapfrog Lemma

Actually, we are going to prove the following even stronger statement. For more information on the background of this theorem we refer the interested reader to [58].

Theorem 11.3.1 *Suppose that* $\mathbf{p}$ *and* $\mathbf{q}$ *are two configurations in* $\mathbb{E}^d, d \geq 1$. *Then the following is a continuous motion* $\mathbf{p}(t) = (\mathbf{p}_1(t), \dots, \mathbf{p}_N(t))$ *in* $\mathbb{E}^{2d}$, *that is analytic in* t, *such that* $\mathbf{p}(0) = \mathbf{p}$, $\mathbf{p}(1) = \mathbf{q}$ *and for* $0 \leq t \leq 1$, $\|\mathbf{p}_i(t) - \mathbf{p}_j(t)\|$ *is monotone:*

$$\mathbf{p}_i(t) = \left(\frac{\mathbf{p}_i + \mathbf{q}_i}{2} + (\cos \pi t) \frac{\mathbf{p}_i - \mathbf{q}_i}{2}, (\sin \pi t) \frac{\mathbf{p}_i - \mathbf{q}_i}{2} \right), \quad 1 \leq i < j \leq N.$$

Proof: We calculate:

$$4\|\mathbf{p}_i(t) - \mathbf{p}_j(t)\|^2 = \|(\mathbf{p}_i - \mathbf{p}_j) - (\mathbf{q}_i - \mathbf{q}_j)\|^2$$

$$+ \|(\mathbf{p}_i - \mathbf{p}_j) + (\mathbf{q}_i - \mathbf{q}_j)\|^2 + 2(\cos \pi t)(\|\mathbf{p}_i - \mathbf{p}_j\|^2 - \|\mathbf{q}_i - \mathbf{q}_j\|^2).$$

This function is monotone, as required.   □

11.4 Proof of Theorem 5.4.1

11.4.1 The spherical leapfrog lemma

As usual, let $\mathbb{S}^d, d \geq 2$ denote the unit sphere centered at the origin $\mathbf{o}$ in $\mathbb{E}^{d+1}$, and let $X(\mathbf{p})$ be a finite intersection of closed balls of radius $\frac{\pi}{2}$ (i.e., of closed hemispheres) in $\mathbb{S}^d$ whose configuration of centers is $\mathbf{p} = (\mathbf{p}_1, \ldots, \mathbf{p}_N)$. We say that another configuration $\mathbf{q} = (\mathbf{q}_1, \ldots, \mathbf{q}_N)$ is a *contraction* of $\mathbf{p}$ if, for all $1 \leq i < j \leq N$, the spherical distance between $\mathbf{p}_i$ and $\mathbf{p}_j$ is not less than the spherical distance between $\mathbf{q}_i$ and $\mathbf{q}_j$. We denote the d-dimensional spherical volume measure by $\mathrm{Svol}_d(\cdot)$. Thus, Theorem 5.4.1, that we need to prove, can be phrased as follows: if $\mathbf{q}$ is a configuration in $\mathbb{S}^d$ that is a contraction of the configuration $\mathbf{p}$, then

$$\mathrm{Svol}_d\left(X(\mathbf{p})\right) \leq \mathrm{Svol}_d\left(X(\mathbf{q})\right). \tag{11.10}$$

We note that the part of Theorem 5.4.1 on the union of closed hemispheres is a simple set-theoretic consequence of (11.10).

Next, we recall Theorem 11.3.1, which we like to call the (Euclidean) Leapfrog Lemma ([58]). We need to apply this to a sphere, rather than Euclidean space. Here we consider the unit spheres $\mathbb{S}^d \subset \mathbb{S}^{d+1} \subset \mathbb{S}^{d+2} \cdots$ in such a way that each $\mathbb{S}^d$ is the set of points that are a unit distance from the origin $\mathbf{o}$ in $\mathbb{E}^{d+1}$. So we need the following.

Corollary 11.4.1 *Suppose that $\mathbf{p}$ and $\mathbf{q}$ are two configurations in $\mathbb{S}^d$. Then there is a monotone analytic motion from $\mathbf{p}$ to $\mathbf{q}$ in $\mathbb{S}^{2d+1}$.*

Proof: Apply Theorem 11.3.1 to each configuration $\mathbf{p}$ and $\mathbf{q}$ with $\mathbf{o}$ as an additional configuration point for each. So for each t, the configuration $\mathbf{p}(t) = (\mathbf{p}_1(t), \ldots, \mathbf{p}_N(t))$ lies at a unit distance from $\mathbf{o}$ in $\mathbb{E}^{2d+2}$, which is just $\mathbb{S}^{2d+1}$. □

11.4.2 Smooth contractions via Schläfli's differential formula

We look at the case when there is a smooth motion $\mathbf{p}(t)$ of the configuration $\mathbf{p}$ in $\mathbb{S}^d$. More precisely we consider the family $X(t) = X(\mathbf{p}(t))$ of convex spherical d-polytopes in $\mathbb{S}^d$ having the same combinatorial face structure with facet hyperplanes being differentiable in the parameter t. The following classical theorem of Schläfli (see, e.g., [182]) describes how the volume of $X(t)$ changes as a function of its dihedral angles and the volume of its $(d-2)$-dimensional faces.

Lemma 11.4.2 *For each $(d-2)$-face $F_{ij}(t)$ of the convex spherical d-polytope $X(t)$ in $\mathbb{S}^d$ let $\alpha_{ij}(t)$ represent the (inner) dihedral angle between the two facets $F_i(t)$ and $F_j(t)$ meeting at $F_{ij}(t)$. Then the following holds.*

$$\frac{d}{dt}\mathrm{Svol}_d\left(X(t)\right) = \frac{1}{d-1}\sum_{F_{ij}}\mathrm{Svol}_{d-2}\left(F_{ij}(t)\right)\cdot\frac{d}{dt}\alpha_{ij}(t),$$

to be summed over all $(d-2)$-faces.

Corollary 11.4.3 *Let* $\mathbf{q}$ *be a configuration in* $\mathbb{S}^d$ *with a differentiable contraction* $\mathbf{p}(t)$ *in* t *of the configuration* $\mathbf{p}$ *in* $\mathbb{S}^d$ *and assume that the convex spherical d-polytopes* $X(t) = X(\mathbf{p}(t))$ *of* $\mathbb{S}^d$ *have the same combinatorial face structure. Then*

$$\frac{d}{dt}\mathrm{Svol}_d\left(X(t)\right) \geq 0 .$$

Proof: As the spherical distance between $\mathbf{p}_i(t)$ and $\mathbf{p}_j(t)$ is decreasing, the derivative of the dihedral angle $\frac{d}{dt}\alpha_{ij}(t) \geq 0$. The result then follows from Lemma 11.4.2. □

11.4.3 Relating higher-dimensional spherical volumes to lower-dimensional ones

The last piece of information that we need before we get to the proof of Theorem 5.4.1 is a way of relating higher-dimensional spherical volumes to lower-dimensional ones. Let X be any integrable set in $\mathbb{S}^n$. Recall that we regard

$$X \subset \mathbb{S}^n = \mathbb{S}^n \times \{\mathbf{o}\} \subset \mathbb{E}^{n+1} \times \mathbb{E}^{k+1} .$$

Regard

$$\{\mathbf{o}\} \times \mathbb{S}^k \subset \mathbb{E}^{n+1} \times \mathbb{E}^{k+1} .$$

Let $X * \mathbb{S}^k$ be the subset of $\mathbb{S}^{n+k+1}$ consisting of the union of the geodesic arcs from each point of X to each point of $\{\mathbf{o}\} \times \mathbb{S}^k$. (So, in particular, $\mathbb{S}^n * \mathbb{S}^k = \mathbb{S}^{n+k+1}$).

Lemma 11.4.4 *For any integrable subset* X *of* $\mathbb{S}^n$,

$$\mathrm{Svol}_{n+k+1}\left(X * \mathbb{S}^k\right) = \frac{\kappa_{n+k+1}}{\kappa_n}\mathrm{Svol}_n(X) ,$$

where $\kappa_n = \mathrm{Svol}_n\left(\mathbb{S}^n\right)$, $\kappa_{n+k+1} = \mathrm{Svol}_{n+k+1}\left(\mathbb{S}^n * \mathbb{S}^k\right) = \mathrm{Svol}_{n+k+1}\left(\mathbb{S}^{n+k+1}\right)$.

Proof: Since the $*$ operation (a kind of spherical join) is associative, we only need to consider the case when $k = 0$. Regard $\{\mathbf{o}\} \times \mathbb{S}^0 = \mathbb{S}^0 = \{\mathbf{n}, \mathbf{s}\}$, the north pole and the south pole of $\mathbb{S}^{n+1}$. We use polar coordinates centered at $\mathbf{n}$ to calculate the $(n+1)$-dimensional volume of $X * \mathbb{S}^0$. Let $X(z) = (X * \mathbb{S}^0) \cap \left(\mathbb{E}^{n+1} \times \{z\}\right)$, and let θ be the angle that a point in $\mathbb{S}^{n+1}$ makes with $\mathbf{n}$, the north pole in $\mathbb{S}^{n+1}$. So $z = z(\theta) = \cos\theta$. Then the spherical volume element for $\mathbb{S}^n(z) = \mathbb{S}^{n+1} \cap \left(\mathbb{E}^{n+1} \times \{z\}\right)$ is $dV_n(z) = (\sin^n\theta)dV_n(0)$ because $\mathbb{S}^n(z)$ is obtained from $\mathbb{S}^n(0)$ by a dilation by $\sin\theta$. Then

$$\text{Svol}_{n+1}\left(X * \mathbb{S}^0\right) = \int_{X * \mathbb{S}^0} dV_n(z)d\theta \tag{11.11}$$

$$= \int_0^\pi \int_{X(z(\theta))} dV_n(z)d\theta = \int_0^\pi (\sin^n \theta)V_n(X)d\theta \tag{11.12}$$

$$= \text{Svol}_n(X) \int_0^\pi (\sin^n \theta)d\theta = \text{Svol}_n(X)\frac{\kappa_{n+1}}{\kappa_n} , \tag{11.13}$$

where (11.13) can be seen by taking $X = \mathbb{S}^n$, or by performing the integral explicitly. $\square$

11.4.4 Putting pieces together

Now, we are ready for the proof of Theorem 5.4.1.

Let the configuration $\mathbf{q} = (\mathbf{q}_1, \ldots, \mathbf{q}_N)$ be a contraction of the configuration $\mathbf{p} = (\mathbf{p}_1, \ldots, \mathbf{p}_N)$ in $\mathbb{S}^d$. By Corollary 11.4.1, there is an analytic motion $\mathbf{p}(t)$, in $\mathbb{S}^{2d+1}$ for $0 \leq t \leq 1$, where $\mathbf{p}(0) = \mathbf{p}$, and $\mathbf{p}(1) = \mathbf{q}$, and all the pairwise distances between the points of $\mathbf{p}(t)$ decrease in t.

Without loss of generality we may assume that $X^d(\mathbf{p}(0)) := X(\mathbf{p}(0))$ is a convex spherical d-polytope in $\mathbb{S}^d$. Since $\mathbf{p}(t)$ is analytic in t, the intersection $X^{2d+1}(\mathbf{p}(t))$ of the (closed) hemispheres centered at the points of the configuration $\mathbf{p}(t)$ in $\mathbb{S}^{2d+1}$ is a convex spherical $(2d+1)$-polytope with a constant combinatorial structure, except for a finite number of points in the interval $[0,1]$. By Corollary 11.4.3, $\text{Svol}_{2d+1}\left(X^{2d+1}(\mathbf{p}(t))\right)$ is monotone increasing in t.

Recall that $X^d(\mathbf{p})$ and $X^d(\mathbf{q})$ are the intersections of the (closed) hemispheres centered at the points of $\mathbf{p}$ and $\mathbf{q}$ in $\mathbb{S}^d$. From the definition of the spherical join $*$,

$$X^d(\mathbf{p}) * \mathbb{S}^d = X^{2d+1}(\mathbf{p}) = X^{2d+1}(\mathbf{p}(0))$$

$$X^d(\mathbf{q}) * \mathbb{S}^d = X^{2d+1}(\mathbf{q}) = X^{2d+1}(\mathbf{p}(1)).$$

Hence, by Lemma 11.4.4,

$$\text{Svol}_d\left(X^d(\mathbf{p})\right) = \frac{\kappa_d}{\kappa_{2d+1}}\text{Svol}_{2d+1}\left(X^{2d+1}(\mathbf{p}(0))\right)$$

$$\leq \frac{\kappa_d}{\kappa_{2d+1}}\text{Svol}_{2d+1}\left(X^{2d+1}(\mathbf{p}(1))\right) = \text{Svol}_d\left(X^d(\mathbf{q})\right) .$$

This finishes the proof of Theorem 5.4.1.

11.5 Proof of Theorem 5.4.6

11.5.1 Monotonicity of the volume of hyperbolic simplices

Case 11.5.1 P *and* **Q** *are simplices.*

Let $\mathbb{X}^n$ be the spherical, Euclidean, or hyperbolic space $\mathbb{S}^n$, $\mathbb{E}^n$, or $\mathbb{H}^n$ of constant curvature $+1$, 0, -1, and of dimension $n \geq 2$. By an n-dimensional simplex Δ^n in $\mathbb{X}^n$ we mean a compact subset with nonempty interior which can be expressed as an intersection of $n+1$ closed halfspaces. (In the case of spherical space we require that Δ^n lies on an open hemisphere.) Let $F_0, F_1, \ldots, F_n$ be the $(n-1)$-dimensional faces of the simplex Δ^n. Each $(n-2)$-dimensional face can be described uniquely as an intersection $F_{ij} = F_i \cap F_j$. We identify the collection of all inner dihedral angles of the simplex Δ^n with the symmetric matrix $\alpha = [\alpha_{ij}]$, where α_{ij} is the inner dihedral angle between F_i and F_j for $i \neq j$, and where the diagonal entries α_{ii} are set equal to π by definition. Then the Gram matrix $G(\Delta^n) = [g_{ij}(\Delta^n)]$ of the simplex $\Delta^n \subset \mathbb{X}^n$ is the $(n+1) \times (n+1)$ symmetric matrix defined by $g_{ij}(\Delta^n) = -\cos \alpha_{ij}$. Note that all diagonal entries $g_{ii}(\Delta^n)$ are equal to one. Finally, let

$$G_+^n := \{G(\Delta^n) \mid \Delta^n \text{ is an } n\text{-dimensional simplex is } \mathbb{S}^n\},$$
$$G_0^n := \{G(\Delta^n) \mid \Delta^n \text{ is an } n\text{-dimensional simplex in } \mathbb{E}^n\},$$
$$G_-^n := \{G(\Delta^n) \mid \Delta^n \text{ is an } n\text{-dimensional simplex in } \mathbb{H}^n\}, \quad \text{and}$$
$$G^n := G_+^n \cup G_0^n \cup G_-^n.$$

The following lemma summarizes some of the major properties of the sets G_+^n, G_0^n, G_-^n, and G^n that have been studied on several occasions including the papers of Coxeter [110], Milnor [197], and Vinberg [244].

Lemma 11.5.2
(1) The determinant of $G(\Delta^n)$ is either positive or zero or negative depending on whether the simplex Δ^n is spherical or Euclidean or hyperbolic.
(2) G^n is an open convex set in $\mathbb{R}^N$ with $N = \frac{n(n+1)}{2}$. (Note that the affine space consisting of all symmetric unidiagonal $(n+1) \times (n+1)$ matrices has dimension $N = \frac{n(n+1)}{2}$.)
(3) G_0^n is an $(N-1)$-dimensional topological cell that cuts G^n into two open subcells G_+^n and G_-^n.
(4) G_+^n (resp., $G_+^n \cup G_0^n$) is an open convex (resp., closed convex) set in $\mathbb{R}^N$.

We need the following property for our proof of Theorem 5.4.6 that seems to be a new property of G_+^n (resp., $G_+^n \cup G_0^n$) not yet mentioned in the literature. It is useful to introduce the notations

$$\mathbb{R}_{<0}^N := \{(x_1, x_2, \ldots, x_N) \mid x_i < 0 \text{ for all } 1 \leq i \leq N\},$$

and

$$\mathbb{R}_{\leq 0}^N := \{(x_1, x_2, \ldots, x_N) \mid x_i \leq 0 \text{ for all } 1 \leq i \leq N\}.$$

Lemma 11.5.3 $G_+^n \cap \mathbb{R}_{<0}^N$ (resp., $(G_+^n \cup G_0^n) \cap \mathbb{R}_{\leq 0}^N$) is a convex corner; that is, if $\mathbf{g} = (g_1, g_2, \ldots, g_N) \in G_+^n \cap \mathbb{R}_{<0}^N$ (resp., $\mathbf{g} \in (G_+^n \cup G_0^n) \cap \mathbb{R}_{\leq 0}^N$), then for any $\mathbf{g}' = (g_1', g_2', \ldots, g_N')$ with $g_1 \leq g_1' < 0, \ldots, g_N \leq g_N' < 0$ (resp., $g_1 \leq g_1' \leq 0, \ldots, g_N \leq g_N' \leq 0$) we have that $\mathbf{g}' \in G_+^n \cap \mathbb{R}_{<0}^N$ (resp. $\mathbf{g}' \in (G_+^n \cup G_0^n) \cap \mathbb{R}_{\leq 0}^N$).

Proof: Due to Lemma 11.5.2 it is sufficient to check the claim of Lemma 11.5.3 for the set $G_+^n \cap \mathbb{R}_{<0}^N$ only. Let $\mathbf{g} = (g_1, g_2, \ldots, g_N) \in G_+^n \cap \mathbb{R}_{<0}^N$. Then it is sufficient to show that for any $\varepsilon_1, \varepsilon_2, \ldots, \varepsilon_N$ with $g_1 \leq \varepsilon_1 < 0, g_2 \leq \varepsilon_2 < 0, \ldots, g_N \leq \varepsilon_N < 0$ we have that

$$\mathbf{g}^1 := (\varepsilon_1, g_2, \ldots, g_N) \in G_+^n \cap \mathbb{R}_{<0}^N,$$
$$\mathbf{g}^2 := (g_1, \varepsilon_2, g_3, \ldots, g_N) \in G_+^n \cap \mathbb{R}_{<0}^N,$$

(*)

$$\vdots$$

$$\mathbf{g}^N := (g_1, \ldots, g_{N-1}, \varepsilon_N) \in G_+^n \cap \mathbb{R}_{<0}^N.$$

Namely, it is easy to see that (*) and the convexity of $G_+^n \cap \mathbb{R}_{<0}^N$ imply that $G_+^n \cap \mathbb{R}_{<0}^N$ is indeed a convex corner. Although it is not needed here, for the sake of completeness we note that the origin of $\mathbb{R}^N$ is in fact, an interior point of G_+^n.

Let Δ^n be the n-dimensional simplex of $\mathbb{S}^n$ whose Gram matrix $G(\Delta^n) = [g_{ij}(\Delta^n)]$ corresponds to $\mathbf{g} = (g_1, g_2, \ldots, g_N)$; that is,

$$(g_1, g_2, \ldots, g_N) = (-\cos\alpha_{01}, \ldots, -\cos\alpha_{0n}, -\cos\alpha_{12}, \ldots, -\cos\alpha_{(n-1)n}).$$

As $\mathbf{g} \in G_+^n \cap \mathbb{R}_{<0}^N$ we have that $0 < \alpha_{01} < \frac{\pi}{2}, 0 < \alpha_{02} < \frac{\pi}{2}, \ldots, 0 < \alpha_{0n} < \frac{\pi}{2}, 0 < \alpha_{12} < \frac{\pi}{2}, \ldots, 0 < \alpha_{(n-1)n} < \frac{\pi}{2}$. In order to show that $\mathbf{g}^1 = (\varepsilon_1, g_2, \ldots g_N) \in G_+^n \cap \mathbb{R}_{<0}^N$ we have to show the existence of an n-dimensional simplex Δ_1^n of $\mathbb{S}^n$ with dihedral angles

$$\arccos(-\varepsilon_1), \alpha_{02}, \ldots, \alpha_{0n}, \alpha_{12}, \ldots, \alpha_{(n-1)n}.$$

(As the task left for the remaining parts of (*) is the same we do not give details of that here.) We show the existence of Δ_1^n via polarity. Let $^*\Delta^n = \{\mathbf{x} \in \mathbb{S}^n \mid \mathbf{x} \cdot \mathbf{y} \leq 0 \text{ for all } \mathbf{y} \in \Delta^n\}$ be the spherical polar of Δ^n, where $\mathbf{x} \cdot \mathbf{y}$ denotes the standard inner product of the unit vectors $\mathbf{x}$ and $\mathbf{y}$. As is well known, $^*\Delta^n$ is an n-dimensional simplex of $\mathbb{S}^n$ with edgelength $\pi - \alpha_{01}, \pi - \alpha_{02}, \ldots, \pi - \alpha_{0n}, \pi - \alpha_{12}, \ldots, \pi - \alpha_{(n-1)n}$ each being larger than $\frac{\pi}{2}$. Let F be the $(n-2)$-dimensional face of $^*\Delta^n$ disjoint from the edge of length $\pi - \alpha_{01}$ of $^*\Delta^n$. Let $\mathbf{v}_0$ and $\mathbf{v}_1$ be the endpoints of the edge of length $\pi - \alpha_{01}$ of $^*\Delta^n$. By assumption $\frac{\pi}{2} < \pi - \arccos(-\varepsilon_1) \leq \pi - \alpha_{01} < \pi$. Now, rotate $\mathbf{v}_1$ towards $\mathbf{v}_0$ about the $(n-2)$-dimensional greatsphere $\mathbb{S}^{n-2}$ of F in $\mathbb{S}^n$ until the rotated image $\bar{\mathbf{v}}_1$ of $\mathbf{v}_1$ becomes a point of the $(n-1)$-dimensional greatsphere $\mathbb{S}^{n-1}$ of the facet of $^*\Delta^n$ disjoint from $\mathbf{v}_1$. Obviously, the above rotation about $\mathbb{S}^{n-2}$ decreases the (spherical) distance $\mathbf{v}_0\mathbf{v}_1$ in a continuous way. We claim via continuity that there is a rotated image say, $\mathbf{v}_{01}$ of $\mathbf{v}_1$ such that the spherical distance $\mathbf{v}_0\mathbf{v}_{01}$ is equal to $\pi - \arccos(-\varepsilon_1)$. Namely, the $n+1$ points formed by $\mathbf{v}_0, \bar{\mathbf{v}}_1$ and the vertices of F all belong to an open hemisphere of $\mathbb{S}^{n-1}$ with the property that all pairwise spherical distances different from $\mathbf{v}_0\bar{\mathbf{v}}_1$ are larger than $\frac{\pi}{2}$. (Here we assume that $\mathbf{v}_0$ and $\bar{\mathbf{v}}_1$ are distinct because if they coincide, then the existence of $\mathbf{v}_{01}$ is trivial.) But, then a theorem of Davenport and Hajós [118] implies that $\mathbf{v}_0\bar{\mathbf{v}}_1 \leq \frac{\pi}{2}$ and so, the existence of $\mathbf{v}_{01}$ follows. Thus, the spherical polar of the n-dimensional simplex of $\mathbb{S}^n$ spanned by $\mathbf{v}_0, \mathbf{v}_{01}$ and F gives us Δ_1^n. This completes the proof of Lemma 11.5.3.   □

Now, we are in a position to show that $G_-^n \cap \mathbb{R}_{\leq 0}^N$ is monotone-path connected.

Lemma 11.5.4 $G_-^n \cap \mathbb{R}_{\leq 0}^N$ *is monotone-path connected in the following strong sense. If* $\mathbf{g} = (g_1, \ldots, g_N) \in G_-^n \cap \mathbb{R}_{\leq 0}^N$ *and* $\mathbf{g}' = (g_1', \ldots, g_N') \in G_-^n \cap \mathbb{R}_{\leq 0}^N$ *with* $g_1' \leq g_1, \ldots, g_N' \leq g_N$, *then* $\lambda\mathbf{g}' + (1-\lambda)\mathbf{g} \in G_-^n \cap \mathbb{R}_{\leq 0}^N$ *for all* $0 \leq \lambda \leq 1$.

Proof: Lemma 11.5.2 implies that $\lambda\mathbf{g}' + (1-\lambda)\mathbf{g} \in G^n$ for all $0 \leq \lambda \leq 1$ and so it is sufficient to prove that $\lambda\mathbf{g}' + (1-\lambda)\mathbf{g} \notin G_+^n \cup G_0^n$ for all $0 \leq \lambda \leq 1$. As $\mathbf{g} \notin G_+^n \cup G_0^n$ and $G_+^n \cup G_0^n$ is convex (Lemma 11.5.2) moreover $(G_+^n \cup G_0^n) \cap \mathbb{R}_{\leq 0}^N$ is a convex corner via Lemma 11.5.3, therefore there exists a supporting hyperplane H in $\mathbb{R}^N$ that touches $G_+^n \cup G_0^n$ at some point $\mathbf{h} \in G_0^n \cap \mathbb{R}_{\leq 0}^N$ and is disjoint from $\mathbf{g}$ and separates $\mathbf{g}$ from $G_+^n \cup G_0^n$. In fact, again using the convex corner property of $(G_+^n \cup G_0^n) \cap \mathbb{R}_{\leq 0}^N$ we get that H separates $\mathbf{h} + \mathbb{R}_{\leq 0}^N$ from $G_+^n \cup G_0^n$ and therefore H separates $\mathbf{g} + \mathbb{R}_{\leq 0}^N$ from $G_+^n \cup G_0^n$ as well. Finally, notice that $\mathbf{g}' \in \mathbf{g} + \mathbb{R}_{\leq 0}^N$ and $\mathbf{g} + \mathbb{R}_{\leq 0}^N$ is disjoint from H and therefore $\mathbf{g} + \mathbb{R}_{\leq 0}^N$ is disjoint from $G_+^n \cup G_0^n$. This finishes the proof of Lemma 11.5.4.   □

Now, we are ready to give a proof of the following volume monotonicity property of hyperbolic simplices.

Theorem 11.5.5 *Let* $\mathbf{P}$ *and* $\mathbf{Q}$ *be nonobtuse-angled n-dimensional hyperbolic simplices. If each inner dihedral angle of* $\mathbf{Q}$ *is at least as large as the corresponding inner dihedral angle of* $\mathbf{P}$, *then the n-dimensional hyperbolic volume of* $\mathbf{P}$ *is at least as large as that of* $\mathbf{Q}$.

Proof: By moving to the space of Gram matrices of n-dimensional hyperbolic simplices and then applying Lemma 11.5.4 we get that there exists a smooth one-parameter family $\mathbf{P}(t)$, $0 \leq t \leq 1$ of nonobtuse-angled n-dimensional hyperbolic simplices with the property that $\mathbf{P}(0) = \mathbf{P}$ and $\mathbf{P}(1) = \mathbf{Q}$; moreover, if $\alpha_{01}(t), \alpha_{02}(t), \ldots, \alpha_{0n}(t), \alpha_{12}(t), \ldots, \alpha_{(n-1)n}(t)$ denote the inner dihedral angles of $\mathbf{P}(t)$, then $\alpha_{ij}(t)$ is a monotone increasing function of t for all $0 \leq i < j \leq n$. Now, Schläfli's classical differential formula (see, e.g., [182]) yields that

$$\frac{d}{dt}\mathrm{Hvol}_n(\mathbf{P}(t)) = \frac{-1}{n-1} \sum_{0 \leq i < j \leq n} \mathrm{Hvol}_{n-2}(F_{ij}(t)) \cdot \frac{d}{dt}\alpha_{ij}(t), \qquad (11.14)$$

where $F_{ij}(t)$ denotes the $(n-2)$-dimensional face of $\mathbf{P}(t)$ on which the dihedral angle $\alpha_{ij}(t)$ sits and $\mathrm{Hvol}_n(\cdot)$, $\mathrm{Hvol}_{n-2}(\cdot)$ refer to the corresponding dimensional hyperbolic volume measures. Thus, as $\frac{d}{dt}\alpha_{ij}(t) \geq 0$, (11.14) implies that $\frac{d}{dt}\mathrm{Hvol}_n(\mathbf{P}(t)) \leq 0$ and so, indeed $\mathrm{Hvol}_n(\mathbf{P}(0)) \geq \mathrm{Hvol}_n(\mathbf{P}(1))$, finishing the proof of Theorem 11.5.5. $\qquad\qquad\qquad\qquad\qquad\qquad\square$

11.5.2 From Andreev's theorem to smooth one-parameter family of hyperbolic polyhedra

Case 11.5.6 *The combinatorial type of* $\mathbf{P}$ *and* $\mathbf{Q}$ *is different from that of a tetrahedron.*

First, recall the following classical theorem of Andreev [6].

Theorem 11.5.7 *A nonobtuse-angled compact convex polyhedron of a given simple combinatorial type, different from that of a tetrahedron and having given inner dihedral angles exists in* $\mathbb{H}^3$ *if and only if the following conditions are satisfied:*
(1) if three faces meet at a vertex, then the sum of the inner dihedral angles between them is larger than π*;*
(2) if three faces are pairwise adjacent but not concurrent, then the sum of the inner dihedral angles between them is smaller than π*;*
(3) if four faces are cyclically adjacent, then at least one of the dihedral angles between them is different from $\frac{\pi}{2}$*;*
(4) (for triangular prism only) one of the angles formed by the lateral faces with the bases must be different from $\frac{\pi}{2}$*.*

Second, observe that the Andreev theorem implies that the space of the inner dihedral angles of nonobtuse-angled compact convex polyhedra of a given simple combinatorial type different from that of a tetrahedron in $\mathbb{H}^3$ is a convex set. As a result we get that if $\mathbf{P}$ and $\mathbf{Q}$ are given as in Theorem 5.4.6 and are different from a tetrahedron, then there exists a smooth one-parameter family $\mathbf{P}(t)$, $0 \leq t \leq 1$ of nonobtuse-angled compact convex polyhedra of the same simple combinatorial type as of $\mathbf{P}$ and $\mathbf{Q}$ with the property that $\mathbf{P}(0) = \mathbf{P}$ and $\mathbf{P}(1) = \mathbf{Q}$; moreover, if $\alpha_E(t)$ denotes the inner dihedral angle of $\mathbf{P}(t)$ which sits over the edge corresponding to the edge E of $\mathbf{P}$, then $\alpha_E(t)$ is a monotone increasing function of t for all edges E of $\mathbf{P}$. Applying Schläfli's differential formula to the smooth one-parameter family $\mathbf{P}(t)$ we get that

$$\frac{d}{dt}\mathrm{Hvol}_3(\mathbf{P}(t)) = -\frac{1}{2}\sum_{E}\mathrm{Hlength}(E_t) \cdot \frac{d}{dt}\alpha_E(t), \qquad (11.15)$$

where E_t denotes the edge of $\mathbf{P}(t)$ corresponding to the edge E of $\mathbf{P}$ and E (resp., E_t) runs over all edges of $\mathbf{P}$ (resp., $\mathbf{P}(t)$). Hence, as $\frac{d}{dt}\alpha_E(t) \geq 0$, (11.15) implies that $\frac{d}{dt}\mathrm{Hvol}_3(\mathbf{P}(t)) \leq 0$ and so, indeed $\mathrm{Hvol}_3(\mathbf{P}(0)) \geq \mathrm{Hvol}_3(\mathbf{P}(1))$, completing the proof of Theorem 5.4.6.

12

Selected Proofs on Ball-Polyhedra

12.1 Proof of Theorem 6.2.1

12.1.1 Finite sets that cannot be translated into the interior of a convex body

We start with the following rather natural statement that can be proved easily with the help of Helly's theorem [85].

Lemma 12.1.1 *Let* $\mathbf{F}$ *be a finite set of at least* $d+1$ *points and* $\mathbf{C}$ *be a convex set in* $\mathbb{E}^d, d \geq 2$*. Then* $\mathbf{C}$ *has a translate that covers* $\mathbf{F}$ *if and only if every* $d+1$ *points of* $\mathbf{F}$ *can be covered by a translate of* $\mathbf{C}$*.*

Proof: For each point $\mathbf{p} \in \mathbf{F}$ let $\mathbf{C_p}$ denote the set of all translation vectors in $\mathbb{E}^d$ with which one can translate $\mathbf{C}$ such that it contains $\mathbf{p}$; that is, let $\mathbf{C_p} := \{\mathbf{t} \in \mathbb{E}^d \mid \mathbf{p} \in \mathbf{t} + \mathbf{C}\}$. Now, it is easy to see that $\mathbf{C_p}$ is a convex set of $\mathbb{E}^d$ for all $\mathbf{p} \in \mathbf{F}$ moreover, $\mathbf{F} \subset \mathbf{t} + \mathbf{C}$ if and only if $\mathbf{t} \in \cap_{\mathbf{p} \in \mathbf{F}} \mathbf{C_p}$. Thus, Helly's theorem [85] applied to the convex sets $\{\mathbf{C_p} \mid \mathbf{p} \in \mathbf{F}\}$ implies that $\mathbf{F} \subset \mathbf{t} + \mathbf{C}$ if and only if $\mathbf{C_{p_1}} \cap \mathbf{C_{p_2}} \cap \cdots \cap \mathbf{C_{p_{d+1}}} \neq \emptyset$ holds for any $\mathbf{p}_1, \mathbf{p}_2, \ldots, \mathbf{p}_{d+1} \in \mathbf{F}$, i.e. if and only if any $\mathbf{p}_1, \mathbf{p}_2, \ldots, \mathbf{p}_{d+1} \in \mathbf{F}$ can be covered by a translate of $\mathbf{C}$, finishing the proof of Lemma 12.1.1. $\square$

Also the following statement plays a central role in our investigations. This is a generalization of the analogue 2-dimensional statement proved in [42].

Lemma 12.1.2 *Let* $\mathbf{F} = \{\mathbf{f}_1, \mathbf{f}_2, \ldots, \mathbf{f}_n\}$ *be a finite set of points and* $\mathbf{C}$ *be a convex body in* $\mathbb{E}^d, d \geq 2$*. Then* $\mathbf{F}$ *cannot be translated into the interior of* $\mathbf{C}$ *if and only if the following two conditions hold. There are closed supporting halfspaces* $H_{i_1}^+, H_{i_2}^+, \ldots, H_{i_s}^+$ *of* $\mathbf{C}$ *assigned to some points of* $\mathbf{F}$ *say, to* $\mathbf{f}_{i_1}, \mathbf{f}_{i_2}, \ldots, \mathbf{f}_{i_s}$ *with* $1 \leq i_1 < i_2 < \cdots < i_s \leq n$ *and a translation vector* $\mathbf{t} \in \mathbb{E}^d$ *such that*
(i) the translated point $\mathbf{t} + \mathbf{f}_{i_j}$ *belongs to the closed halfspace* $H_{i_j}^-$ *for all* $1 \leq j \leq s$*, where the interior of* $H_{i_j}^-$ *is disjoint from the interior of* $H_{i_j}^+$ *and its*

boundary hyperplane is identical to the boundary hyperplane of $H_{i_j}^+$ (which is in fact, a supporting hyperplane of $\mathbf{C}$);
(ii) the intersection $\cap_{j=1}^s H_{i_j}^+$ is nearly bounded, meaning that it lies between two parallel hyperplanes of $\mathbb{E}^d$.

Proof: First, we assume that there are closed supporting halfspaces $H_{i_1}^+$, $H_{i_2}^+, \ldots, H_{i_s}^+$ of $\mathbf{C}$ assigned to some points of $\mathbf{F}$ say, to $\mathbf{f}_{i_1}, \mathbf{f}_{i_2}, \ldots, \mathbf{f}_{i_s}$ with $1 \leq i_1 < i_2 < \cdots < i_s \leq n$ and a translation vector $\mathbf{t} \in \mathbb{E}^d$ satisfying (i) as well as (ii). Based on this our goal is to show that $\mathbf{F}$ cannot be translated into the interior of $\mathbf{C}$ or equivalently that $\mathbf{F}$ cannot be covered by a translate of the interior int$\mathbf{C}$ of $\mathbf{C}$. We prove this in an indirect way: we assume that $\mathbf{F}$ can be covered by a translate of int$\mathbf{C}$ and look for a contradiction. Indeed, if $\mathbf{F}$ can be covered by a translate of int$\mathbf{C}$, then $\mathbf{t} + \mathbf{F}$ can be covered by a translate of int$\mathbf{C}$; that is, there is a translation vector $\mathbf{t}^* \in \mathbb{E}^d$ such that $\mathbf{t} + \mathbf{F} \subset \mathbf{t}^* + \text{int}\mathbf{C}$. In particular, if $\mathbf{F}^* := \{\mathbf{f}_{i_1}, \mathbf{f}_{i_2}, \ldots, \mathbf{f}_{i_s}\}$, then $\mathbf{t} + \mathbf{F}^* \subset \mathbf{t}^* + \text{int}\mathbf{C}$. Clearly, this implies that $\cap_{j=1}^s H_{i_j}^+ \subset \text{int}\left(\cap_{j=1}^s \mathbf{t}^* + H_{i_j}^+\right)$, a contradiction to (ii).

Second, we assume that $\mathbf{F}$ cannot be translated into the interior of $\mathbf{C}$ and look for closed supporting halfspaces $H_{i_1}^+, H_{i_2}^+, \ldots, H_{i_s}^+$ of $\mathbf{C}$ assigned to some points of $\mathbf{F}$ say, to $\mathbf{f}_{i_1}, \mathbf{f}_{i_2}, \ldots, \mathbf{f}_{i_s}$ with $1 \leq i_1 < i_2 < \cdots < i_s \leq n$ and a translation vector $\mathbf{t} \in \mathbb{E}^d$ satisfying (i) as well as (ii). In order to simplify matters let us start to investigate the case when $\mathbf{C}$ is a smooth convex body in $\mathbb{E}^d$, that is, when through each boundary point of $\mathbf{C}$ there exists precisely one supporting hyperplane of $\mathbf{C}$. (Also, without loss of generality we assume that the origin $\mathbf{o}$ of $\mathbb{E}^d$ is an interior point of $\mathbf{C}$.) As $\mathbf{F}$ cannot be translated into int$\mathbf{C}$ therefore Lemma 12.1.1 implies that there are $m \leq d+1$ points of $\mathbf{F}$ say, $\mathbf{F}_m := \{\mathbf{f}_{j_1}, \mathbf{f}_{j_2}, \ldots, \mathbf{f}_{j_m}\}$ with $1 \leq j_1 < j_2 < \cdots < j_m \leq n$ such that $\mathbf{F}_m$ cannot be translated into int$\mathbf{C}$. Now, let $\lambda_0 := \inf\{\lambda > 0 \mid \lambda\mathbf{F}_m$ cannot be translated into int$\mathbf{C}\}$. Clearly, $\lambda_0 \leq 1$ and $\lambda_0\mathbf{F}_m$ cannot be translated into int$\mathbf{C}$; moreover, as $\lambda_0 = \sup\{\delta > 0 \mid \delta\mathbf{F}_m$ can be translated into $\mathbf{C}\}$, therefore there exists a translation vector $\mathbf{t} \in \mathbb{E}^d$ such that $\mathbf{t} + \lambda_0\mathbf{F}_m \subset \mathbf{C}$. Let $\mathbf{t} + \lambda_0\mathbf{f}_{i_1}, \mathbf{t} + \lambda_0\mathbf{f}_{i_2}, \ldots, \mathbf{t} + \lambda_0\mathbf{f}_{i_s}$ with $1 \leq i_1 < i_2 < \cdots < i_s \leq n, 2 \leq s \leq m \leq d+1$ denote the points of $\mathbf{t} + \lambda_0\mathbf{F}_m$ that are boundary points of $\mathbf{C}$ and let $H_{i_1}^+, H_{i_2}^+, \ldots, H_{i_s}^+$ be the corresponding closed supporting halfspaces of $\mathbf{C}$. We claim that $\mathbf{H}^+ := \cap_{k=1}^s H_{i_k}^+$ is nearly bounded. Indeed, if $\mathbf{H}^+$ were not nearly bounded, then there would be a translation vector $\mathbf{t}' \in \mathbb{E}^d$ with $\mathbf{H}^+ \subset \mathbf{t}' + \text{int}\mathbf{H}^+$. As $\mathbf{C}$ is a smooth convex body therefore this would imply the existence of a sufficiently small $\mu > 0$ with the property that $\{\mathbf{t} + \lambda_0\mathbf{f}_{i_1}, \mathbf{t} + \lambda_0\mathbf{f}_{i_2}, \ldots, \mathbf{t} + \lambda_0\mathbf{f}_{i_s}\} \subset \mu\mathbf{t}' + \text{int}\mathbf{C}$, a contradiction. Thus, as $\mathbf{o} \in \text{int}\mathbf{C}$ therefore the points $\mathbf{f}_{i_1}, \mathbf{f}_{i_2}, \ldots, \mathbf{f}_{i_s}$ and the closed supporting halfspaces $H_{i_1}^+, H_{i_2}^+, \ldots, H_{i_s}^+$ and the translation vector $\mathbf{t} \in \mathbb{E}^d$ satisfy (i) as well as (ii). We are left with the case when $\mathbf{C}$ is not necessarily a smooth convex body in $\mathbb{E}^d$. In this case let $\mathbf{C}_N, N = 1, 2, \ldots$ be a sequence of smooth convex bodies lying in int$\mathbf{C}$ with $\lim_{N \to +\infty} \mathbf{C}_N = \mathbf{C}$. As $\mathbf{F}$ cannot be translated into the interior of $\mathbf{C}_N$ for all $N = 1, 2, \ldots$ therefore applying the method described above to each $\mathbf{C}_N$ and taking proper subsequences if necessary we end up with some points of $\mathbf{F}$ say, $\mathbf{f}_{i_1}, \mathbf{f}_{i_2}, \ldots, \mathbf{f}_{i_s}$ with $1 \leq i_1 < i_2 < \cdots < i_s \leq n$ and with s convergent sequences of closed supporting halfspaces $H^+_{N,i_1}, H^+_{N,i_2}, \ldots, H^+_{N,i_s}$ of $\mathbf{C}_N$ and a convergent sequence of translation vectors $\mathbf{t}_N$ that satisfy (i) and (ii) for each N. By taking the limits $H^+_{i_1} := \lim_{N \to +\infty} H^+_{N,i_1}, H^+_{i_2} := \lim_{N \to +\infty} H^+_{N,i_2}, \ldots, H^+_{i_s} := \lim_{N \to +\infty} H^+_{N,i_s}$, and $\mathbf{t} := \lim_{N \to +\infty} \mathbf{t}_N$ we get the desired nearly bounded family of closed supporting halfspaces of $\mathbf{C}$ and the translation vector $\mathbf{t} \in \mathbb{E}^d$ satisfying (i) as well as (ii). This completes the proof of Lemma 12.1.2. □

12.1.2 From generalized billiard trajectories to shortest ones

Lemma 12.1.3 *Let $\mathbf{C}$ be a convex body in $\mathbb{E}^d, d \geq 2$. If $\mathbf{P}$ is a generalized billiard trajectory in $\mathbf{C}$, then $\mathbf{P}$ cannot be translated into the interior of $\mathbf{C}$.*

Proof: Let $\mathbf{p}_1, \mathbf{p}_2, \ldots, \mathbf{p}_n$ be the vertices of $\mathbf{P}$ and let $\mathbf{v}_1, \mathbf{v}_2, \ldots, \mathbf{v}_n$ be the points of the unit sphere $\mathbb{S}^{d-1}$ centered at the origin $\mathbf{o}$ in $\mathbb{E}^d$ whose position vectors are parallel to the inner angle bisectors (halflines) of $\mathbf{P}$ at the vertices $\mathbf{p}_1, \mathbf{p}_2, \ldots, \mathbf{p}_n$ of $\mathbf{P}$. Moreover, let $H^+_1, H^+_2, \ldots, H^+_n$ denote the closed supporting halfspaces of $\mathbf{C}$ whose boundary hyperplanes are perpendicular to the inner angle bisectors of $\mathbf{P}$ at the vertices $\mathbf{p}_1, \mathbf{p}_2, \ldots, \mathbf{p}_n$. Based on Lemma 12.1.2 in order to prove that $\mathbf{P}$ cannot be translated into the interior of $\mathbf{C}$ it is sufficient to show that $\cap_{i=1}^n H^+_i$ is nearly bounded or equivalently that $\mathbf{o} \in \text{conv}(\{\mathbf{v}_1, \mathbf{v}_2, \ldots, \mathbf{v}_n\})$, where conv(.) denotes the convex hull of the corresponding set in $\mathbb{E}^d$. It is easy to see that $\mathbf{o} \in \text{conv}(\{\mathbf{v}_1, \mathbf{v}_2, \ldots, \mathbf{v}_n\})$ if and only if for any hyperplane H of $\mathbb{E}^d$ passing through $\mathbf{o}$ and for any of the two closed halfspaces bounded by H say, for H^+, we have that $H^+ \cap \text{conv}(\{\mathbf{v}_1, \mathbf{v}_2, \ldots, \mathbf{v}_n\}) \neq \emptyset$. Indeed, for a given H^+ let $\mathbf{t} \in \mathbb{E}^d$ be chosen so that $\mathbf{t} + H^+$ is a supporting halfspace of $\text{conv}(\{\mathbf{p}_1, \mathbf{p}_2, \ldots, \mathbf{p}_n\})$. Clearly, at least one vertex of $\mathbf{P}$ say, $\mathbf{p}_{i_0}$ must belong to the boundary of $\mathbf{t} + H^+$ and therefore $\mathbf{v}_{i_0} \in H^+ \cap \text{conv}(\{\mathbf{v}_1, \mathbf{v}_2, \ldots, \mathbf{v}_n\})$, finishing the proof of Lemma 12.1.3. □

For the purpose of the following statement it seems natural to introduce *generalized $(d + 1)$-gons* in $\mathbb{E}^d$ as closed polygonal paths (possibly with self-intersections) having at most $d + 1$ sides.

Theorem 12.1.4 *Let $\mathbf{C}$ be a convex body in $\mathbb{E}^d, d \geq 2$ and let $\mathcal{F}_{d+1}(\mathbf{C})$ denote the family of all generalized $(d+1)$-gons of $\mathbb{E}^d$ that cannot be translated into the interior of $\mathbf{C}$. Then $\mathcal{F}_{d+1}(\mathbf{C})$ possesses a minimal length member; moreover, the shortest perimeter members of $\mathcal{F}_{d+1}(\mathbf{C})$ are identical (up to translations) with the shortest generalized billiard trajectories of $\mathbf{C}$.*

Proof: If $\mathbf{P}$ is an arbitrary generalized billiard trajectory of the convex body $\mathbf{C}$ in $\mathbb{E}^d$ with vertices $\mathbf{p}_1, \mathbf{p}_2, \ldots, \mathbf{p}_n$, then according to Lemma 12.1.3 $\mathbf{P}$ cannot be translated into the interior of $\mathbf{C}$. Thus, by Lemma 12.1.1 $\mathbf{P}$ possesses at most $d + 1$ vertices say, $\mathbf{p}_{i_1}, \mathbf{p}_{i_2}, \ldots, \mathbf{p}_{i_{d+1}}$ with $1 \leq i_1 \leq i_2 \leq \cdots \leq i_{d+1} \leq n$

such that $\mathbf{p}_{i_1}, \mathbf{p}_{i_2}, \ldots, \mathbf{p}_{i_{d+1}}$ cannot be translated into the interior of $\mathbf{C}$. This implies that by connecting the consecutive points of $\mathbf{p}_{i_1}, \mathbf{p}_{i_2}, \ldots, \mathbf{p}_{i_{d+1}}$ by line segments according to their cyclic ordering the generalized $(d + 1)$-gon $\mathbf{P}_{d+1}$ obtained, has length $l(\mathbf{P}_{d+1})$ at most as large as the length $l(\mathbf{P})$ of $\mathbf{P}$; moreover, $\mathbf{P}_{d+1}$ cannot be covered by a translate of int$\mathbf{C}$ (i.e., $\mathbf{P}_{d+1} \in \mathcal{F}_{d+1}(\mathbf{C})$). Now, by looking at only those members of $\mathcal{F}_{d+1}(\mathbf{C})$ that lie in a d-dimensional ball of sufficiently large radius in $\mathbb{E}^d$ we get via a standard compactness argument and Lemma 12.1.2 that $\mathcal{F}_{d+1}(\mathbf{C})$ possesses a member of minimal length say, $\varDelta_{d+1}(\mathbf{C})$. As the inequalities $l(\varDelta_{d+1}(\mathbf{C})) \leq l(\mathbf{P}_{d+1}) \leq l(\mathbf{P})$ hold for any generalized billiard trajectory $\mathbf{P}$ of $\mathbf{C}$, therefore in order to finish our proof it is sufficient to show that $\varDelta_{d+1}(\mathbf{C})$ is a generalized billiard trajectory of $\mathbf{C}$. Indeed, as $\varDelta_{d+1}(\mathbf{C}) \in \mathcal{F}_{d+1}(\mathbf{C})$ therefore $\varDelta_{d+1}(\mathbf{C})$ cannot be translated into int$\mathbf{C}$. Thus, the minimality of $\varDelta_{d+1}(\mathbf{C})$ and Lemma 12.1.2 imply that if $\mathbf{q}_1, \mathbf{q}_2, \ldots, \mathbf{q}_m$ denote the vertices of $\varDelta_{d+1}(\mathbf{C})$ with $m \leq d + 1$, then there are closed supporting halfspaces $H_1^+, H_2^+, \ldots, H_m^+$ of $\mathbf{C}$ whose boundary hyperplanes $H_1, H_2, \ldots, H_m$ pass through the points $\mathbf{q}_1, \mathbf{q}_2, \ldots, \mathbf{q}_m$ (each being a boundary point of $\mathbf{C}$) and have the property that $\cap_{i=1}^m H_i^+$ is nearly bounded in $\mathbb{E}^d$. If the inner angle bisector at a vertex of $\varDelta_{d+1}(\mathbf{C})$ say, at $\mathbf{q}_i$ were not perpendicular to H_i, then it is easy to see via Lemma 12.1.2 that one could slightly move $\mathbf{q}_i$ along H_i to a new position $\mathbf{q}_i'$ (which is typically an exterior point of $\mathbf{C}$ on H_i) such that the new generalized $(d+1)$-gon $\varDelta_{d+1}'(\mathbf{C}) \in \mathcal{F}_{d+1}(\mathbf{C})$ would have a shorter length, a contradiction. This completes the proof of Lemma 12.1.4.   □

Finally, notice that Theorem 6.2.1 follows from Theorem 12.1.4 in a straightforward way.

12.2 Proofs of Theorems 6.6.1, 6.6.3, and 6.6.4

12.2.1 Strict separation by spheres of radii at most one

For the proof of Theorem 6.6.1 we need the following weaker version of it due to Houle [169] as well as the following lemma proved in [69].

Theorem 12.2.1 *Let $A, B \subset \mathbb{E}^d$ be finite sets. Then A and B can be strictly separated by a sphere $S^{d-1}(\mathbf{c}, r)$ such that $A \subset \mathbf{B}^d(\mathbf{c}, r)$ if and only if for every $T \subset A \cup B$ with card$T \leq d + 2$, $T \cap A$ and $T \cap B$ can be strictly separated by a sphere $S^{d-1}(\mathbf{c}_T, r_T)$ such that $T \cap A \subset \mathbf{B}^d(\mathbf{c}_T, r_T)$.*

Lemma 12.2.2 *Let $A, B \subset \mathbb{E}^d$ be finite sets and suppose that $S^{d-1}(\mathbf{o}, 1)$ is the smallest sphere that separates A from B such that $A \subset \mathbf{B}^d[\mathbf{o}, 1]$. Then there is a set $T \subset A \cup B$ with card$T \leq d + 1$ such that $S^{d-1}(\mathbf{o}, 1)$ is the smallest sphere $S^{d-1}(\mathbf{c}, r)$ that separates $T \cap A$ from $T \cap B$ and satisfies $T \cap A \subset \mathbf{B}^d[\mathbf{c}, r]$.*

We prove the "if" part of Theorem 6.6.1; the opposite direction is trivial. Theorem 12.2.1 guarantees the existence of the smallest sphere $S^{d-1}(\mathbf{c}', r')$

that separates A and B such that $A \subset \mathbf{B}^d[\mathbf{c}', r']$. According to Lemma 12.2.2, there is a set $T \subset A \cup B$ with card$T \leq d+1$ such that $S^{d-1}(\mathbf{c}', r')$ is the smallest sphere that separates $T \cap A$ from $T \cap B$ and whose convex hull contains $T \cap A$. By the assumption, we have $r' < r_T \leq 1$. Note that Theorem 12.2.1 guarantees the existence of a sphere $S^{d-1}(\mathbf{c}^*, r^*)$ that strictly separates A from B and satisfies $A \subset \mathbf{B}^d(\mathbf{c}^*, r^*)$. Because $r' < 1$, there is a sphere $S^{d-1}(\mathbf{c}, r)$ with $r \leq 1$ such that $\mathbf{B}^d[\mathbf{c}', r'] \cap \mathbf{B}^d(\mathbf{c}^*, r^*) \subset \mathbf{B}^d(\mathbf{c}, r) \subset \mathbb{E}^d \setminus (\mathbf{B}^d(\mathbf{c}', r') \cup \mathbf{B}^d[\mathbf{c}^*, r^*])$. This sphere clearly satisfies the conditions in Theorem 6.6.1 and so, the proof of Theorem 6.6.1 is complete.

12.2.2 Characterizing spindle convex sets

Our proof of Theorem 6.6.3 is based on the following statement.

Lemma 12.2.3 *Let a spindle convex set $\mathbf{C} \subset \mathbb{E}^d$ be supported by the hyperplane H in $\mathbb{E}^d$ at $\mathbf{x} \in \mathrm{bd}\mathbf{C}$. Then the closed unit ball supported by H at $\mathbf{x}$ and lying in the same side as $\mathbf{C}$ contains $\mathbf{C}$.*

Proof: Let $\mathbf{B}^d[\mathbf{c}, 1]$ be the closed unit ball that is supported by H at $\mathbf{x}$ and is in the same closed half-space bounded by H as $\mathbf{C}$. We show that $\mathbf{B}^d[\mathbf{c}, 1]$ is the desired unit ball.

Assume that $\mathbf{C}$ is not contained in $\mathbf{B}^d[\mathbf{c}, 1]$. So, there is a point $\mathbf{y} \in \mathbf{C}$, $y \notin \mathbf{B}^d[\mathbf{c}, 1]$. Then, by taking the intersection of the configuration with the plane that contains $\mathbf{x}, \mathbf{y}$, and $\mathbf{c}$, we see that there is a shorter unit circular arc connecting $\mathbf{x}$ and $\mathbf{y}$ that does not intersect $\mathbf{B}^d(\mathbf{c}, 1)$. Hence, H cannot be a supporting hyperplane of $\mathbf{C}$ at $\mathbf{x}$, a contradiction.   □

Indeed, it is easy to see that Lemma 12.2.3 implies Theorem 6.6.3 in a rather straightforward way.

12.2.3 Separating spindle convex sets

Finally, we prove Theorem 6.6.4 as follows. Since $\mathbf{C}$ and $\mathbf{D}$ are spindle convex, they are convex bounded sets with disjoint relative interiors. So, their closures are convex compact sets with disjoint relative interiors. Hence, they can be separated by a hyperplane H that supports $\mathbf{C}$ at a point, say $\mathbf{x}$. The closed unit ball $\mathbf{B}^d[\mathbf{c}, 1]$ of Lemma 12.2.3 satisfies the conditions of the first statement of Theorem 6.6.4. For the second statement of Theorem 6.6.4, we assume that $\mathbf{C}$ and $\mathbf{D}$ have disjoint closures, so $\mathbf{B}^d[\mathbf{c}, 1]$ is disjoint from the closure of $\mathbf{D}$ and remains so even after a sufficiently small translation. Furthermore, $\mathbf{C}$ is a spindle convex set that is different from a unit ball, so $\mathbf{c} \notin \mathrm{conv}(\mathbf{C} \cap S^{d-1}(\mathbf{c}, 1))$. Hence, there is a sufficiently small translation of $\mathbf{B}^d[\mathbf{c}, 1]$ that satisfies the second statement of Theorem 6.6.4, finishing the proof of Theorem 6.6.4.

12.3 Proof of Theorem 6.7.1

12.3.1 On the boundary of spindle convex hulls in terms of supporting spheres

Let $S^k(\mathbf{c}, r) \subset \mathbb{E}^d$ be a k-dimensional sphere centered at $\mathbf{c}$ and having radius r with $0 \leq k \leq d-1$. Recall the following strong version of spherical convexity. A set $F \subset S^k(\mathbf{c}, r)$ is *spherically convex* if it is contained in an open hemisphere of $S^k(\mathbf{c}, r)$ and for every $\mathbf{x}, \mathbf{y} \in F$ the shorter great-circular arc of $S^k(\mathbf{c}, r)$ connecting $\mathbf{x}$ with $\mathbf{y}$ is in F. The *spherical convex hull* of a set $X \subset S^k(\mathbf{c}, r)$ is defined in the natural way and it exists if and only if X is in an open hemisphere of $S^k(\mathbf{c}, r)$. We denote it by $\mathrm{Sconv}(X, S^k(\mathbf{c}, r))$. Carathéodory's theorem can be stated for the sphere in the following way. If $X \subset S^k(\mathbf{c}, r)$ is a set in an open hemisphere of $S^k(\mathbf{c}, r)$, then $\mathrm{Sconv}(X, S^k(\mathbf{c}, r))$ is the union of spherical simplices with vertices in X. The proof of this spherical equivalent of the original Carathéodory's theorem uses the central projection of the open hemisphere of $S^k(\mathbf{c}, r)$ to $\mathbb{E}^k$.

Recall that the circumradius $\mathrm{cr}(X)$ of a bounded set $X \subset \mathbb{E}^d$ is defined as the radius of the unique smallest d-dimensional closed ball that contains X (also known as the circumball of X). Now, it is easy to see that if $C \subset \mathbb{E}^d$ is a spindle convex set such that $C \subset \mathbf{B}^d[\mathbf{q}, 1]$ and $\mathrm{cr}(C) < 1$, then $C \cap S^{d-1}(\mathbf{q}, 1)$ is spherically convex on $S^{d-1}(\mathbf{q}, 1)$.

The following lemma describes the surface of a spindle convex hull.

Lemma 12.3.1 *Let $X \subset \mathbb{E}^d$ be a closed set such that $\mathrm{cr}(X) < 1$ and let $\mathbf{B}^d[\mathbf{q}, 1]$ be a closed unit ball containing X. Then*
(i) $X \cap S^{d-1}(\mathbf{q}, 1)$ is contained in an open hemisphere of $S^{d-1}(\mathbf{q}, 1)$,
(ii) $\mathrm{conv}_s(X) \cap S^{d-1}(\mathbf{q}, 1) = \mathrm{Sconv}(X \cap S^{d-1}(\mathbf{q}, 1), S^{d-1}(\mathbf{q}, 1))$.

Proof: Because $\mathrm{cr}(X) < 1$, we obtain that X is contained in the intersection of two distinct closed unit balls which proves (i). Note that by (i), the right-hand side $Z := \mathrm{Sconv}(X \cap S^{d-1}(\mathbf{q}, 1), S^{d-1}(\mathbf{q}, 1))$ of (ii) exists. We show that the set on the left-hand side is contained in Z; the other containment follows from the discussion right before Lemma 12.3.1.

Suppose that $\mathbf{y} \in \mathrm{conv}_s(X) \cap S^{d-1}(\mathbf{q}, 1)$ is not contained in Z. We show that there is a hyperplane H through $\mathbf{q}$ that strictly separates Z from $\mathbf{y}$. Consider an open hemisphere of $S^{d-1}(\mathbf{q}, 1)$ that contains Z, call the spherical center of this hemisphere $\mathbf{p}$. If $\mathbf{y}$ is an exterior point of the hemisphere, H exists. If $\mathbf{y}$ is on the boundary of the hemisphere, then, by moving the hemisphere a little, we find another open hemisphere that contains Z, but with respect to which $\mathbf{y}$ is an exterior point.

Assume that $\mathbf{y}$ is contained in the open hemisphere. Let L be a hyperplane tangent to $S^{d-1}(\mathbf{q}, 1)$ at p. We project Z and $\mathbf{y}$ centrally from $\mathbf{q}$ onto L and, by the separation theorem of convex sets in L, we obtain a $(d-2)$-dimensional affine subspace T of L that strictly separates the image of Z from the image of $\mathbf{y}$. Then $H := \mathrm{aff}(T \cup \{\mathbf{q}\})$ is the desired hyperplane.

Hence, $\mathbf{y}$ is contained in one open hemisphere of $S^{d-1}(\mathbf{q}, 1)$ and Z is in the other. Let $\mathbf{v}$ be the unit normal vector of H pointing towards the hemisphere of $S^{d-1}(\mathbf{q}, 1)$ that contains Z. Since X is closed, its distance from the closed hemisphere containing $\mathbf{y}$ is positive. Hence, we can move $\mathbf{q}$ a little in the direction $\mathbf{v}$ to obtain the point $\mathbf{q}'$ such that $X \subset \mathbf{B}^d[\mathbf{q}, 1] \cap \mathbf{B}^d[\mathbf{q}', 1]$ and $\mathbf{y} \notin \mathbf{B}^d[\mathbf{q}', 1]$. As $\mathbf{B}^d[\mathbf{q}', 1]$ separates X from $\mathbf{y}$, the latter is not in $\mathrm{conv}_s(X)$, a contradiction. $\square$

12.3.2 From the spherical Carathéodory theorem to an analogue for spindle convex hulls

Now, we prove Theorem 6.7.1.

Assume that $\mathrm{cr}(X) > 1$. Recall that the intersection of the d-dimensional closed unit balls of $\mathbb{E}^d$ centered at the points of X is denoted by $\mathbf{B}[X]$. Then $\mathbf{B}[X] = \emptyset$; hence, by Helly's theorem, there is a set $\{\mathbf{x}_0, \mathbf{x}_1, \ldots, \mathbf{x}_d\} \subset X$ such that $\mathbf{B}[\{\mathbf{x}_0, \mathbf{x}_1, \ldots, \mathbf{x}_d\}] = \emptyset$. It follows that $\mathrm{conv}_s(\{\mathbf{x}_0, \mathbf{x}_1, \ldots, \mathbf{x}_d\}) = \mathbb{E}^d$. Thus, (i) and (ii) follow.

Now, we prove (i) for $\mathrm{cr}(X) < 1$. By the spherical Carathéodory theorem, Lemma 12.2.3, and Lemma 12.3.1 we obtain that

$$\mathbf{y} \in \mathrm{Sconv}(\{\mathbf{x}_1, \mathbf{x}_2, \ldots, \mathbf{x}_d\}, S^{d-1}(\mathbf{q}, 1))$$

for some $\{\mathbf{x}_1, \mathbf{x}_2, \ldots, \mathbf{x}_d\} \subset X$ and some $\mathbf{q} \in \mathbb{E}^d$ such that $X \subset \mathbf{B}^d[\mathbf{q}, 1]$. Hence, $\mathbf{y} \in \mathrm{conv}_s\{\mathbf{x}_1, \mathbf{x}_2, \ldots, \mathbf{x}_d\}$.

We prove (i) for $\mathrm{cr}(X) = 1$ by a limit argument as follows. Without loss of generality, we may assume that $X \subset \mathbf{B}^d[\mathbf{o}, 1]$. Let $X^k := (1 - \frac{1}{k})X$ for any $k \in \mathbb{Z}^+$. Let $\mathbf{y}^k$ be the point of $\mathrm{bd}\,(\mathrm{conv}_s(X^k))$ closest to $\mathbf{y}$. Thus, $\lim_{k \to \infty} \mathbf{y}^k = \mathbf{y}$. Clearly, $\mathrm{cr}(X^k) < 1$, hence there is a set $\{\mathbf{x}_1^k, \mathbf{x}_2^k, \ldots, \mathbf{x}_d^k\} \subset X^k$ such that $\mathbf{y}^k \in \mathrm{conv}_s\{\mathbf{x}_1^k, \mathbf{x}_2^k, \ldots, \mathbf{x}_d^k\}$. By compactness, there is a sequence $0 < i_1 < i_2 < \cdots$ of indices such that all the d sequences $\{\mathbf{x}_1^{i_j} : j \in \mathbb{Z}^+\}, \{\mathbf{x}_2^{i_j} : j \in \mathbb{Z}^+\}, \ldots, \{\mathbf{x}_d^{i_j} : j \in \mathbb{Z}^+\}$ converge. Let their respective limits be $\mathbf{x}_1, \mathbf{x}_2, \ldots, \mathbf{x}_d$. Since X is closed, these d points are contained in X. Clearly, $\mathbf{y} \in \mathrm{conv}_s\{\mathbf{x}_1, \mathbf{x}_2, \ldots, \mathbf{x}_d\}$.

To prove (ii) for $\mathrm{cr}(X) \leq 1$, suppose that $\mathbf{y} \in \mathrm{int}\,(\mathrm{conv}_s X)$. Then let $\mathbf{x}_0 \in X \cap \mathrm{bd}\,(\mathrm{conv}_s X)$ be arbitrary and let $\mathbf{y}_1$ be the intersection of $\mathrm{bd}\,(\mathrm{conv}_s X)$ with the ray starting from $\mathbf{x}_0$ and passing through $\mathbf{y}$. Now, by (i), $\mathbf{y}_1 \in \mathrm{conv}_s\{\mathbf{x}_1, \mathbf{x}_2, \ldots, \mathbf{x}_d\}$ for some $\{\mathbf{x}_1, \mathbf{x}_2, \ldots, \mathbf{x}_d\} \subset X$. Then clearly $\mathbf{y} \in \mathrm{int}\,(\mathrm{conv}_s\{\mathbf{x}_0, \mathbf{x}_1, \ldots, \mathbf{x}_d\})$.

12.4 Proof of Theorem 6.8.3

12.4.1 On the boundary of spindle convex hulls in terms of normal images

Let $X \subset \mathbb{E}^d, d \geq 3$ be a compact set of Euclidean diameter $\mathrm{diam}(X) \leq 1$. Recall that $\mathbf{B}[X] \subset \mathbb{E}^d$ denotes the convex body which is the intersection of the closed unit balls of $\mathbb{E}^d$ centered at the points of X. For the following investigations it is more proper to use the normal images than the Gauss images of the boundary points of $\mathbf{B}[X]$ defined as follows. The *normal image* $N_{\mathbf{B}[X]}(\mathbf{b})$ of the boundary point $\mathbf{b} \in \mathrm{bd}\,(\mathbf{B}[X])$ of $\mathbf{B}[X]$ is

$$N_{\mathbf{B}[X]}(\mathbf{b}) := -\nu(\{\mathbf{b}\})$$

In other words, $N_{\mathbf{B}[X]}(\mathbf{b}) \subset \mathbb{S}^{d-1}$ is the set of inward unit normal vectors of all hyperplanes that support $\mathbf{B}[X]$ at $\mathbf{b}$. Clearly, $N_{\mathbf{B}[X]}(\mathbf{b})$ is a closed spherically convex subset of $\mathbb{S}^{d-1}$. (Here we refer to the strong version of spherical convexity introduced for Lemma 12.3.1.)

We need to introduce the following notation as follows. For a set $A \subset \mathbb{S}^{d-1}$ let $A^+ = \{\mathbf{x} \in \mathbb{S}^{d-1} \mid \langle \mathbf{x}, \mathbf{y} \rangle > 0 \text{ for all } \mathbf{y} \in A\}$. (Here $\| \cdot \|$ and $\langle \cdot, \cdot \rangle$ refer to the canonical Euclidean norm and the canonical inner product on $\mathbb{E}^d$.)

As is well known, illumination can be reformulated as follows: *The direction* $\mathbf{u} \in \mathbb{S}^{d-1}$ *illuminates the boundary point* $\mathbf{b}$ *of the convex body* $\mathbf{B}[X]$ *if and only if* $\mathbf{u} \in N_{\mathbf{B}[X]}(\mathbf{b})^+$. (Because the proof of this claim is straightforward we leave it to the reader. For more insight on illumination we refer the interested reader to [47] and the relevant references listed there.)

Finally, we need to recall some further notations as well. Let $\mathbf{a}$ and $\mathbf{b}$ be two points in $\mathbb{E}^d$. If $\|\mathbf{a} - \mathbf{b}\| < 2$, then the *(closed) spindle* of $\mathbf{a}$ and $\mathbf{b}$, denoted by $[\mathbf{a}, \mathbf{b}]_s$, is defined as the union of circular arcs with endpoints $\mathbf{a}$ and $\mathbf{b}$ that are of radii at least one and are shorter than a semicircle. If $\|\mathbf{a} - \mathbf{b}\| = 2$, then $[\mathbf{a}, \mathbf{b}]_s := \mathbf{B}^d[\frac{\mathbf{a}+\mathbf{b}}{2}, 1]$, where $\mathbf{B}^d[\mathbf{p}, r]$ denotes the (closed) d-dimensional ball centered at $\mathbf{p}$ with radius r in $\mathbb{E}^d$. If $\|\mathbf{a} - \mathbf{b}\| > 2$, then we define $[\mathbf{a}, \mathbf{b}]_s$ to be $\mathbb{E}^d$. Next, a set $\mathbf{C} \subset \mathbb{E}^d$ is called *spindle convex* if, for any pair of points $\mathbf{a}, \mathbf{b} \in \mathbf{C}$, we have that $[\mathbf{a}, \mathbf{b}]_s \subset \mathbf{C}$. Finally, let X be a set in $\mathbb{E}^d$. Then the *spindle convex hull* of X is the set defined by $\mathrm{conv}_s X := \bigcap\{C \subset \mathbb{E}^d \mid X \subset C \text{ and } C \text{ is spindle convex in } \mathbb{E}^d\}$.

Now, we are ready to state Lemma 12.4.1, which is the core part of this section and whose proof is based on Lemma 12.3.1.

Lemma 12.4.1 *Let* $X \subset \mathbb{E}^d, d \geq 3$ *be a compact set of Euclidean diameter* $\mathrm{diam}(X) \leq 1$. *Then the boundary of the spindle convex hull of* X *can be generated as follows:*

$$\mathrm{bd}\,(\mathrm{conv}_s(X)) = \bigcup_{\mathbf{b} \in \mathrm{bd}(\mathbf{B}[X])} \{\mathbf{b} + \mathbf{y} \mid \mathbf{y} \in N_{\mathbf{B}[X]}(\mathbf{b})\}.$$

Proof: Let $\mathbf{b} \in \mathrm{bd}\,(\mathbf{B}[X])$. Then (ii) of Lemma 12.3.1 implies that

$$\mathbf{b} + N_{\mathbf{B}[X]}(\mathbf{b}) = \mathrm{Sconv}(X \cap S^{d-1}(\mathbf{b}, 1), S^{d-1}(\mathbf{b}, 1)) = \mathrm{conv}_s(X) \cap S^{d-1}(\mathbf{b}, 1).$$

This together with the fact that

$$\bigcup_{\mathbf{b} \in \mathrm{bd}(\mathbf{B}[X])} N_{\mathbf{B}[X]}(\mathbf{b}) = \mathbb{S}^{d-1}$$

finishes the proof of Lemma 12.4.1.   □

12.4.2 On the Euclidean diameter of spindle convex hulls and normal images

Lemma 12.4.2

$$\mathrm{diam}\,(\mathrm{conv}_s(X)) \leq 1.$$

Proof: By assumption $\mathrm{diam}(X) \leq 1$. Recall that Meissner [196] has called a compact set $M \subset \mathbb{E}^d$ *complete* if $\mathrm{diam}(M \cup \{\mathbf{p}\}) > \mathrm{diam}(M)$ for any $\mathbf{p} \in \mathbb{E}^d \setminus M$. He has proved in [196] that any set of diameter 1 is contained in a complete set of diameter 1. Moreover, he has shown in [196] that a compact set of diameter 1 in $\mathbb{E}^d$ is complete if and only if it is of constant width 1. These facts together with the easy observation that any convex body of constant width 1 in $\mathbb{E}^d$ is in fact a spindle convex set, imply that X is contained in a convex body of convex width 1 and any such convex body must necessarily contain $\mathrm{conv}_s(X)$. Thus, indeed $\mathrm{diam}\,(\mathrm{conv}_s(X)) \leq 1$.   □

For an arbitrary nonempty subset A of $\mathbb{S}^{d-1}$ let

$$U_{\mathbf{B}[X]}(A) = \left(\bigcup_{N_{\mathbf{B}[X]}(\mathbf{b}) \cap A \neq \emptyset} N_{\mathbf{B}[X]}(\mathbf{b}) \right) \subset \mathbb{S}^{d-1}.$$

Lemma 12.4.3 *Let $\emptyset \neq A \subset \mathbb{S}^{d-1}$ be given. Then*

$$\mathrm{diam}\,(U_{\mathbf{B}[X]}(A)) \leq 1 + \mathrm{diam}(A).$$

Proof: Let $\mathbf{y}_1 \in N_{\mathbf{B}[X]}(\mathbf{b}_1)$ and $\mathbf{y}_2 \in N_{\mathbf{B}[X]}(\mathbf{b}_2)$ be two arbitrary points of $U_{\mathbf{B}[X]}(A)$ with $\mathbf{b}_1, \mathbf{b}_2 \in \mathrm{bd}\,(\mathbf{B}[X])$. We need to show that $\|\mathbf{y}_1 - \mathbf{y}_2\| \leq 1 + \mathrm{diam}(A)$.

By Lemma 12.4.1 and by Lemma 12.4.2 we get that

$$\|(\mathbf{y}_1 - \mathbf{y}_2) + (\mathbf{b}_1 - \mathbf{b}_2)\| = \|(\mathbf{b}_1 + \mathbf{y}_1) - (\mathbf{b}_2 + \mathbf{y}_2)\| \leq 1.$$

Thus, the triangle inequality yields that

$$\|(\mathbf{y}_1 - \mathbf{y}_2)\| \leq 1 + \|(\mathbf{b}_2 - \mathbf{b}_1)\|.$$

This means that in order to finish the proof of Lemma 12.4.3 it is sufficient to show that $\|(\mathbf{b}_2 - \mathbf{b}_1)\| \leq \operatorname{diam}(A)$. This can be obtained easily from the assumption that $N_{\mathbf{B}[X]}(\mathbf{b}_1) \cap A \neq \emptyset$, $N_{\mathbf{B}[X]}(\mathbf{b}_2) \cap A \neq \emptyset$ and from the fact that the sets $\mathbf{b}_1 + N_{\mathbf{B}[X]}(\mathbf{b}_1) \subset \operatorname{bd}(\operatorname{conv}_s(X))$ and $\mathbf{b}_2 + N_{\mathbf{B}[X]}(\mathbf{b}_2) \subset \operatorname{bd}(\operatorname{conv}_s(X))$ are separated by the hyperplane H of $\mathbb{E}^d$ that bisects the line segment connecting $\mathbf{b}_1$ to $\mathbf{b}_2$ and is perpendicular to it with $\mathbf{b}_1 + N_{\mathbf{B}[X]}(\mathbf{b}_1)$ (resp., $\mathbf{b}_2 + N_{\mathbf{B}[X]}(\mathbf{b}_2)$) lying on the same side of H as $\mathbf{b}_2$ (resp., $\mathbf{b}_1$). □

12.4.3 An upper bound for the illumination number based on a probabilistic approach

Let μ_{d-1} denote the standard probability measure on $\mathbb{S}^{d-1}$ and define

$$V_{d-1}(t) := \inf\{\mu_{d-1}(A^+) \mid A \subset \mathbb{S}^{d-1}, \operatorname{diam}(A) \leq t\},$$

where just as before $A^+ = \{\mathbf{x} \in \mathbb{S}^{d-1} \mid \langle \mathbf{x}, \mathbf{y} \rangle > 0 \text{ for all } \mathbf{y} \in A\}$. Moreover, let $n_{d-1}(\epsilon)$ denote the minimum number of closed spherical caps of $\mathbb{S}^{d-1}$ having Euclidean diameter ϵ such that they cover $\mathbb{S}^{d-1}$, where $0 < \epsilon \leq 2$.

Lemma 12.4.4

$$I(\mathbf{B}[X]) \leq 1 + \frac{\ln(n_{d-1}(\epsilon))}{-\ln(1 - V_{d-1}(1+\epsilon))}$$

holds for all $0 < \epsilon \leq \sqrt{2} - 1$ and $d \geq 3$.

Proof: Let $\emptyset \neq A \subset \mathbb{S}^{d-1}$ be given with Euclidean diameter $\operatorname{diam}(A) \leq 1 + \epsilon \leq \sqrt{2}$. Then the spherical Jung theorem [119] implies that A is contained in a closed spherical cap of $\mathbb{S}^{d-1}$ having angular radius $0 < \arcsin\sqrt{\frac{d-1}{d}} < \frac{\pi}{2}$. Thus, A^+ contains a spherical cap of $\mathbb{S}^{d-1}$ having angular radius $\frac{\pi}{2} - \arcsin\sqrt{\frac{d-1}{d}} > 0$ and of course, A^+ is contained in an open hemisphere of $\mathbb{S}^{d-1}$. Hence, $0 < V_{d-1}(1+\epsilon) < \frac{1}{2}$ and so, the expression on the right in Lemma 12.4.4 is well defined.

Let m be a positive integer satisfying

$$m > \frac{\ln(n_{d-1}(\epsilon))}{-\ln(1 - V_{d-1}(1+\epsilon))}.$$

It is sufficient to show that m directions can illuminate $\mathbf{B}[X]$. Let $n = n_{d-1}(\epsilon)$ and let $A_1, A_2, \ldots, A_n$ be closed spherical caps of $\mathbb{S}^{d-1}$ having Euclidean diameter ϵ and covering $\mathbb{S}^{d-1}$. By Lemma 12.4.3 we have $\operatorname{diam}\left(U_{\mathbf{B}[X]}(A_i)\right) \leq 1 + \epsilon$ for all $1 \leq i \leq n$ and therefore

$$\mu_{d-1}\left(U_{\mathbf{B}[X]}(A_i)^+\right) \geq V_{d-1}(1+\epsilon)$$

for all $1 \leq i \leq n$. Let the directions $\mathbf{u}_1, \mathbf{u}_2, \ldots, \mathbf{u}_m$ be chosen at random, uniformly and independently distributed on $\mathbb{S}^{d-1}$. Thus, the probability that

$\mathbf{u}_j$ lies in $U_{\mathbf{B}[X]}(A_i)^+$ is equal to $\mu_{d-1}\left(U_{\mathbf{B}[X]}(A_i)^+\right) \geq V_{d-1}(1+\epsilon)$. Therefore the probabilty that $U_{\mathbf{B}[X]}(A_i)^+$ contains none of the points $\mathbf{u}_1, \mathbf{u}_2, \ldots, \mathbf{u}_m$ is at most $(1 - V_{d-1}(1+\epsilon))^m$. Hence, the probability p that at least one $U_{\mathbf{B}[X]}(A_i)^+$ will contain none of the points $\mathbf{u}_1, \mathbf{u}_2, \ldots, \mathbf{u}_m$ satisfies

$$p \leq \sum_{i=1}^{n}(1 - V_{d-1}(1+\epsilon))^m < n\,(1 - V_{d-1}(1+\epsilon))^{\frac{\ln(n)}{-\ln(1-V_{d-1}(1+\epsilon))}} = 1.$$

This shows that one can choose m directions say, $\{\mathbf{v}_1, \mathbf{v}_2, \ldots, \mathbf{v}_m\} \subset \mathbb{S}^{d-1}$ such that each set $U_{\mathbf{B}[X]}(A_i)^+, 1 \leq i \leq n$ contains at least one of them. We claim that the directions $\mathbf{v}_1, \mathbf{v}_2, \ldots, \mathbf{v}_m$ illuminate $\mathbf{B}[X]$. Indeed, let $\mathbf{b} \in \mathrm{bd}\,(\mathbf{B}[X])$. We show that at least one of the directions $\mathbf{v}_1, \mathbf{v}_2, \ldots, \mathbf{v}_m$ illuminates the boundary point $\mathbf{b}$. As the spherical caps $A_1, A_2, \ldots, A_n$ form a covering of $\mathbb{S}^{d-1}$ therefore there exists an A_i with $A_i \cap N_{\mathbf{B}[X]}(\mathbf{b}) \neq \emptyset$. Thus, by definition $N_{\mathbf{B}[X]}(\mathbf{b}) \subset U_{\mathbf{B}[X]}(A_i)$ and therefore

$$N_{\mathbf{B}[X]}(\mathbf{b})^+ \supset U_{\mathbf{B}[X]}(A_i)^+.$$

$U_{\mathbf{B}[X]}(A_i)^+$ contains at least one of the directions $\mathbf{v}_1, \mathbf{v}_2, \ldots, \mathbf{v}_m$, say $\mathbf{v}_k$. Hence,

$$\mathbf{v}_k \in U_{\mathbf{B}[X]}(A_i)^+ \subset N_{\mathbf{B}[X]}(\mathbf{b})^+$$

and so, $\mathbf{v}_k$ illuminates the boundary point $\mathbf{b}$ of $\mathbf{B}[X]$, finishing the proof of Lemma 12.4.4. $\qquad\square$

12.4.4 Schramm's lower bound for the proper measure of polars of sets of given diameter in spherical space

We need the following notation for the next statement. For $\mathbf{u} \in \mathbb{S}^{d-1}$ let $R_{\mathbf{u}} : \mathbb{E}^d \to \mathbb{E}^d$ denote the reflection about the line passing through the points $\mathbf{u}$ and $-\mathbf{u}$. Clearly, $R_{\mathbf{u}}(\mathbf{x}) = 2\langle \mathbf{x}, \mathbf{u}\rangle\mathbf{u} - \mathbf{x}$ for all $\mathbf{x} \in \mathbb{E}^d$.

Lemma 12.4.5 *Let* $A \subset \mathbb{S}^{d-1}$ *be a set of Euclidean diameter* $0 < \mathrm{diam}(A) \leq t$ *contained in the closed spherical cap* $C[\mathbf{u}, \arccos a] \subset \mathbb{S}^{d-1}$ *centered at* $\mathbf{u} \in \mathbb{S}^{d-1}$ *having angular radius* $0 < \arccos a < \frac{\pi}{2}$ *with* $0 < a < 1$ *and* $0 < t \leq 2\sqrt{1-a^2}$. *Then*

$$A^+ \cup R_{\mathbf{u}}(A^+) \supset C\left(\mathbf{u}, \arctan\left(\frac{2a}{t}\right)\right),$$

where $C\left(\mathbf{u}, \arctan\left(\frac{2a}{t}\right)\right) \subset \mathbb{S}^{d-1}$ *denotes the open spherical cap centered at* $\mathbf{u}$ *having angular radius* $0 < \arctan(\frac{2a}{t}) < \frac{\pi}{2}$.

Proof: Suppose that $\mathbf{x} \in \mathbb{S}^{d-1} \setminus (A^+ \cup R_{\mathbf{u}}(A^+))$ and let θ denote the angular distance between $\mathbf{x}$ and $\mathbf{u}$. Clearly $0 < \theta \leq \pi$ and

$$\mathbf{x} = (\cos\theta)\mathbf{u} + (\sin\theta)\mathbf{v}$$

with $\mathbf{v} \in \mathbb{S}^{d-1}$ being perpendicular to $\mathbf{u}$. As $\mathbf{x} \notin A^+$ (resp., $\mathbf{x} \notin R_\mathbf{u}(A^+)$ i.e. $R_\mathbf{u}(\mathbf{x}) \notin A^+$) therefore there exists a point $\mathbf{y} \in A$ (resp., $\mathbf{z} \in A$) such that

$$0 \geq \langle \mathbf{y}, \mathbf{u} \rangle \cos \theta + \langle \mathbf{y}, \mathbf{v} \rangle \sin \theta \text{ (resp., } 0 \geq \langle \mathbf{z}, \mathbf{u} \rangle \cos \theta - \langle \mathbf{z}, \mathbf{v} \rangle \sin \theta).$$

By adding together the last two inequalities and using the inequalities $\|\mathbf{y} - \mathbf{z}\| \leq t$ and $\sin \theta \geq 0$ we get that

$$0 \geq \langle \mathbf{y} + \mathbf{z}, \mathbf{u} \rangle \cos \theta + \langle \mathbf{y} - \mathbf{z}, \mathbf{v} \rangle \sin \theta \geq \langle \mathbf{y} + \mathbf{z}, \mathbf{u} \rangle \cos \theta - t \sin \theta.$$

As $A \subset C[\mathbf{u}, \arccos a] \subset \mathbb{S}^{d-1}$ therefore if $\cos \theta > 0$, then the last inequality implies that

$$\tan \theta \geq \frac{\langle \mathbf{y} + \mathbf{z}, \mathbf{u} \rangle}{t} = \frac{\langle \mathbf{y}, \mathbf{u} \rangle + \langle \mathbf{z}, \mathbf{u} \rangle}{t} \geq \frac{2a}{t}.$$

Thus, $\theta \geq \arctan \left(\frac{2a}{t} \right)$ follows for all $0 < \theta \leq \pi$, finishing the proof of Lemma 12.4.5. □

Lemma 12.4.6

$$V_{d-1}(t) \geq \frac{1}{\sqrt{8\pi d}} \left(\frac{3}{2} + \frac{\left(2 - \frac{1}{d} \right) t^2 - 2}{4 - \left(2 - \frac{2}{d} \right) t^2} \right)^{-\frac{d-1}{2}}$$

for all $0 < t < \sqrt{\frac{2d}{d-1}}$ and $d \geq 3$.

Proof: Let $\emptyset \neq A \subset \mathbb{S}^{d-1}$ be given with (Euclidean) diameter $\mathrm{diam}(A) \leq t$. The spherical Jung theorem [119] implies that A is contained in the closed spherical cap $C \left[\mathbf{u}, \arcsin \left(\sqrt{\frac{d-1}{2d}} t \right) \right] \subset \mathbb{S}^{d-1}$ centered at the properly chosen $\mathbf{u} \in \mathbb{S}^{d-1}$ having angular radius $0 < \arcsin \left(\sqrt{\frac{d-1}{2d}} t \right) < \frac{\pi}{2}$, where by assumption $0 < t < \sqrt{\frac{2d}{d-1}}$. Thus, Lemma 12.4.5 implies that

$$A^+ \cup R_\mathbf{u}(A^+) \supset C \left(\mathbf{u}, \arctan \left(\frac{2a}{t} \right) \right)$$

with $a = \sqrt{1 - \frac{d-1}{2d} t^2}$. Hence,

$$\mu_{d-1}(A^+) = \frac{1}{2} \left(\mu_{d-1}(A^+) + \mu_{d-1}(R_\mathbf{u}(A^+)) \right) \geq \frac{1}{2} \mu_{d-1} \left(A^+ \cup R_\mathbf{u}(A^+) \right)$$

$$\geq \frac{1}{2} \mu_{d-1} \left(C \left(\mathbf{u}, \arctan \left(\frac{2a}{t} \right) \right) \right) = \frac{1}{2} \frac{\mathrm{Svol}_{d-1} \left(C \left(\mathbf{u}, \arctan \left(\frac{2a}{t} \right) \right) \right)}{\mathrm{Svol}_{d-1}(\mathbb{S}^{d-1})}$$

$$= \frac{\mathrm{Svol}_{d-1}\left(C\left(\mathbf{u}, \arctan\left(\frac{2a}{t}\right)\right)\right)}{2d\omega_d} = \frac{\mathrm{Svol}_{d-1}\left(C\left[\mathbf{u}, \arctan\left(\frac{2a}{t}\right)\right]\right)}{2d\omega_d}.$$

As $\sin\left(\arctan\left(\frac{2a}{t}\right)\right) = \left(1+\frac{t^2}{4a^2}\right)^{-\frac{1}{2}}$ therefore

$$\mathrm{Svol}_{d-1}\left(C\left[\mathbf{u}, \arctan\left(\frac{2a}{t}\right)\right]\right)$$

$$> \mathrm{vol}_{d-1}\left(\mathbf{B}^{d-1}\left[\cos\left(\arctan\left(\frac{2a}{t}\right)\right)\mathbf{u}, \left(1+\frac{t^2}{4a^2}\right)^{-\frac{1}{2}}\right]\right)$$

$$= \left(1+\frac{t^2}{4a^2}\right)^{-\frac{d-1}{2}}\omega_{d-1} \text{ and so, } \mu_{d-1}(A^+) \geq \frac{\omega_{d-1}}{2d\omega_d}\left(1+\frac{t^2}{4a^2}\right)^{-\frac{d-1}{2}}.$$

Hence, using the well-known estimate (see also [226]) $\frac{\omega_{d-1}}{\omega_d} \geq \sqrt{\frac{d}{2\pi}}$ we get that

$$\mu_{d-1}(A^+) \geq \frac{1}{2d}\sqrt{\frac{d}{2\pi}}\left(1+\frac{t^2}{4a^2}\right)^{-\frac{d-1}{2}}.$$

Finally, substituting $a = \sqrt{1-\frac{d-1}{2d}t^2}$ we are led to the following inequality

$$\mu_{d-1}(A^+) \geq \frac{1}{\sqrt{8\pi d}}\left(1+\frac{t^2}{4-\frac{2(d-1)t^2}{d}}\right)^{-\frac{d-1}{2}}$$

$$= \frac{1}{\sqrt{8\pi d}}\left(\frac{3}{2}+\frac{\left(2-\frac{1}{d}\right)t^2-2}{4-\left(2-\frac{2}{d}\right)t^2}\right)^{-\frac{d-1}{2}}.$$

This finishes the proof of Lemma 12.4.6.   □

12.4.5 An upper bound for the number of sets of given diameter that are needed to cover spherical space

Lemma 12.4.7

$$n_{d-1}(\epsilon) < \left(1+\frac{4}{\epsilon}\right)^d$$

for all $0 < \epsilon \leq 2$ and $d \geq 3$.

Proof: Let $\{\mathbf{p}_1, \mathbf{p}_2, \ldots \mathbf{p}_n\} \subset \mathbb{S}^{d-1}$ be the largest family of points on $\mathbb{S}^{d-1}$ with the property that $\|\mathbf{p}_i - \mathbf{p}_j\| \geq \frac{\epsilon}{2}$ for all $1 \leq i < j \leq n$. Then clearly $\bigcup_{i=1}^n \mathbf{B}^d\left[\mathbf{p}_i, \frac{\epsilon}{2}\right] \supset \mathbb{S}^{d-1}$ and therefore $n \geq n_{d-1}(\epsilon)$. As the balls $\mathbf{B}^d[\mathbf{p}_i, \frac{\epsilon}{4}], 1 \leq i \leq n$ form a packing in $\mathbf{B}^d[\mathbf{o}, 1+\frac{\epsilon}{4}]$ therefore

$$n\left(\frac{\epsilon}{4}\right)^d \omega_d < \left(1+\frac{\epsilon}{4}\right)^d \omega_d,$$

implying that

$$n_{d-1}(\epsilon) \le n < \frac{\left(1+\frac{\epsilon}{4}\right)^d}{\left(\frac{\epsilon}{4}\right)^d} = \left(1+\frac{4}{\epsilon}\right)^d.$$

This completes the proof of Lemma 12.4.7.   □

Actually, using [122], one can replace the inequality of Lemma 12.4.7 by the stronger inequality $n_{d-1}(\epsilon) \le (\frac{1}{2}+o(1))d\ln d\left(\frac{2}{\epsilon}\right)^d$. As this improves the estimate of Theorem 6.8.3 only in a rather insignificant way, we do not introduce it here.

12.4.6 The final upper bound for the illumination number

Now, we are ready for the proof of Theorem 6.8.3. As $x < -\ln(1-x)$ holds for all $0 < x < 1$, therefore by Lemma 12.4.4 we get that

$$I(\mathbf{B}[X]) \le 1 + \frac{\ln\left(n_{d-1}(\epsilon)\right)}{-\ln\left(1 - V_{d-1}(1+\epsilon)\right)} < 1 + \frac{\ln\left(n_{d-1}(\epsilon)\right)}{V_{d-1}(1+\epsilon)}$$

holds for all $0 < \epsilon \le \sqrt{2}-1$ and $d \ge 3$. Now, let $\epsilon_0 = \sqrt{\frac{2d}{2d-1}} - 1$. As $0 < \epsilon_0 < \sqrt{2}-1$ holds for all $d \ge 3$, therefore Lemma 12.4.6 and Lemma 12.4.7 together with the easy inequality $\epsilon_0 > \frac{4}{16d-1}$ yield that

$$I(\mathbf{B}[X]) < 1 + \sqrt{8\pi d}\left(\frac{3}{2}\right)^{\frac{d-1}{2}}\ln\left(n_{d-1}(\epsilon_0)\right)$$

$$< 1 + \sqrt{8\pi d}\left(\frac{3}{2}\right)^{\frac{d-1}{2}}\ln\left(\left(1+\frac{4}{\epsilon_0}\right)^d\right) < 1 + \sqrt{8\pi d}\left(\frac{3}{2}\right)^{\frac{d-1}{2}}\ln\left((16d)^d\right)$$

$$= 1 + 4\sqrt{\frac{\pi}{3}d}\sqrt{d}\left(\frac{3}{2}\right)^{\frac{d}{2}}(\ln 16 + \ln d) < 4\left(\frac{\pi}{3}\right)^{\frac{1}{2}}d^{\frac{3}{2}}(3+\ln d)\left(\frac{3}{2}\right)^{\frac{d}{2}},$$

finishing the proof of Theorem 6.8.3.

12.5 Proof of Theorem 6.9.1

12.5.1 The CW-decomposition of the boundary of a standard ball-polyhedron

Let $\mathbf{K}$ be a convex body in $\mathbb{E}^d$ and $\mathbf{b} \in \mathrm{bd}\mathbf{K}$. Then recall that the Gauss image of $\mathbf{b}$ with respect to $\mathbf{K}$ is the set of outward unit normal vectors of hyperplanes that support $\mathbf{K}$ at $\mathbf{b}$. Clearly, it is a spherically convex subset of $S^{d-1}(\mathbf{o}, 1)$ and its dimension is defined in the natural way.

Theorem 12.5.1 *Let* **P** *be a standard ball-polyhedron. Then the faces of* **P** *form the closed cells of a finite CW-decomposition of the boundary of* **P**.

Proof: Let $\{S^{d-1}(\mathbf{p}_1, 1), \ldots, S^{d-1}(\mathbf{p}_k, 1)\}$ be the reduced family of generating spheres of **P**. The relative interior (resp., the relative boundary) of an m-dimensional face F of **P** is defined as the set of those points of F that are mapped to $\mathbf{B}^m(\mathbf{o}, 1)$ (resp., $S^{m-1}(\mathbf{o}, 1)$) under any homeomorphism between F and $\mathbf{B}^m[\mathbf{o}, 1]$. For every $\mathbf{b} \in \mathrm{bd}\mathbf{P}$ define the following sphere

$$S(\mathbf{b}) := \bigcap \{S^{d-1}(\mathbf{p}_i, 1) : \mathbf{p}_i \in S^{d-1}(\mathbf{b}, 1), i \in \{1, \ldots, k\}\}.$$

Clearly, $S(\mathbf{b})$ is a support sphere of **P**. Moreover, if $S(\mathbf{b})$ is an m-dimensional sphere, then the face $F := S(\mathbf{b}) \cap \mathbf{P}$ is also m-dimensional as $\mathbf{b}$ has an m-dimensional neighbourhood in $S(\mathbf{b})$ that is contained in F. This also shows that $\mathbf{b}$ belongs to the relative interior of F. Hence, the union of the relative interiors of the faces covers $\mathrm{bd}\mathbf{P}$.

We claim that every face F of **P** can be obtained in this way; that is, for any relative interior point $\mathbf{b}$ of F we have $F = S(\mathbf{b}) \cap \mathbf{P}$. Clearly, $F \supset S(\mathbf{b}) \cap \mathbf{P}$, as the support sphere of **P** that intersects **P** in F contains $S(\mathbf{b})$. It is sufficient to show that F is at most m-dimensional. This is so, because the Gauss image of $\mathbf{b}$ with respect to **P** is at least $(d - m - 1)$-dimensional, since the Gauss image of $\mathbf{b}$ with respect to $\bigcap \{\mathbf{B}^d[\mathbf{p}_i, 1] : \mathbf{p}_i \in S^{d-1}(\mathbf{b}, 1), i \in \{1, \ldots, k\}\} \supset \mathbf{P}$ is $(d - m - 1)$-dimensional.

The above argument also shows that no point $\mathbf{b} \in \mathrm{bd}\mathbf{P}$ belongs to the relative interior of more than one face. Moreover, if $\mathbf{b} \in \mathrm{bd}\mathbf{P}$ is on the relative boundary of the face F then $S(\mathbf{b})$ is clearly of smaller dimension than F. Hence, $\mathbf{b}$ belongs to the relative interior of a face of smaller dimension. This concludes the proof of Theorem 12.5.1. □

12.5.2 On the number of generating balls of a standard ball-polyhedron

Corollary 12.5.2 *The generating balls of any standard ball-polyhedron* **P** *in* $\mathbb{E}^d$ *consist of at least $d + 1$ unit balls.*

Proof: because the faces form a CW-decomposition of the boundary of **P**, there is a vertex **v**. The Gauss image of **v** is $(d - 1)$-dimensional. So, **v** belongs to at least d generating spheres from the family of generating balls. We denote the centers of those spheres by $\mathbf{x}_1, \mathbf{x}_2, \ldots, \mathbf{x}_d$. Let $H := \mathrm{aff}\{\mathbf{x}_1, \mathbf{x}_2, \ldots, \mathbf{x}_d\}$. Then $\mathbf{B}[\{\mathbf{x}_1, \mathbf{x}_2, \ldots, \mathbf{x}_d\}]$, which denotes the intersection of the closed d-dimensional unit balls centered at the points $\mathbf{x}_1, \mathbf{x}_2, \ldots, \mathbf{x}_d$, is symmetric about H. Let σ_H be the reflection of $\mathbb{E}^d$ about H. Then $S := S^{d-1}(\mathbf{x}_1, 1) \cap S^{d-1}(\mathbf{x}_2, 1) \cap \cdots \cap S^{d-1}(\mathbf{x}_d, 1)$ contains the points **v** and $\sigma_H(\mathbf{v})$, hence S is a sphere, not a point. Finally, as **P** is a standard ball-polyhedron, therefore there is a unit-ball $\mathbf{B}^d[\mathbf{x}_{d+1}, 1]$ in the family of generating balls of **P** that does not contain S. □

12.5.3 Basic properties of face lattices of standard ball-polyhedra

Corollary 12.5.3 *Let Λ be the set containing all faces of a standard ball-polyhedron $\mathbf{P} \subset \mathbb{E}^d$ and the empty set and $\mathbf{P}$ itself. Then Λ is a finite bounded lattice with respect to ordering by inclusion. The atoms of Λ are the vertices of $\mathbf{P}$ and Λ is atomic: for every element $F \in \Lambda$ with $F \neq \emptyset$ there is a vertex $\mathbf{x}$ of $\mathbf{P}$ such that $\mathbf{x} \in F$.*

Proof: First, we show that the intersection of two faces F_1 and F_2 is another face (or the empty set). The intersection of the two supporting spheres that intersect $\mathbf{P}$ in F_1 and F_2 is another supporting sphere of $\mathbf{P}$, say $S^l(\mathbf{p}, r)$. Then $S^l(\mathbf{p}, r) \cap \mathbf{P} = F_1 \cap F_2$ is a face of $\mathbf{P}$. From this the existence of a unique maximum common lower bound (i.e., an infimum) for F_1 and F_2 follows.

Moreover, by the finiteness of Λ, the existence of a unique infimum for any two elements of Λ implies the existence of a unique minimum common upper bound (i.e., a supremum) for any two elements of Λ, say C and D, as follows. The supremum of C and D is the infimum of all the (finitely many) elements of Λ that are above C and D.

Vertices of $\mathbf{P}$ are clearly atoms of Λ. Using Theorem 12.5.1 and induction on the dimension of the face it is easy to show that every face is the supremum of its vertices. $\quad\square$

Corollary 12.5.4 *A standard ball-polyhedron $\mathbf{P}$ in $\mathbb{E}^d$ has k-dimensional faces for every $0 \leq k \leq d - 1$.*

Proof: We use an inductive argument on k, where we go from $k = d-1$ down to $k = 0$. Clearly, $\mathbf{P}$ has facets. A k-face F of $\mathbf{P}$ is homeomorphic to $\mathbf{B}^k[\mathbf{o}, 1]$, hence its relative boundary is homeomorphic to $S^{k-1}(\mathbf{o}, 1)$, if $k > 0$. Since the $(k-1)$-skeleton of $\mathbf{P}$ covers the relative boundary of F, $\mathbf{P}$ has $(k-1)$-faces. $\square$

Corollary 12.5.5 *Let $d \geq 3$. Any standard ball-polyhedron $\mathbf{P}$ is the spindle convex hull of its $(d - 2)$-dimensional faces. Furthermore, no standard ball-polyhedron is the spindle convex hull of its $(d - 3)$-dimensional faces.*

Proof: For the first statement, it is sufficient to show that the spindle convex hull of the $(d - 2)$-faces contains the facets. Let $\mathbf{p}$ be a point on the facet, $F = \mathbf{P} \cap S^{d-1}(\mathbf{q}, 1)$. Take any great circle C of $S^{d-1}(\mathbf{q}, 1)$ passing through $\mathbf{p}$. Since F is spherically convex on $S^{d-1}(\mathbf{q}, 1)$, $C \cap F$ is a unit circular arc of length less than π. Let $\mathbf{r}, \mathbf{s} \in S^{d-1}(\mathbf{q}, 1)$ be the two endpoints of $C \cap F$. Then $\mathbf{r}$ and $\mathbf{s}$ belong to the relative boundary of F. Hence, by Theorem 12.5.1, $\mathbf{r}$ (resp., $\mathbf{s}$) belongs to a $(d - 2)$-face. Clearly, $\mathbf{p} \in \operatorname{conv}_s\{\mathbf{r}, \mathbf{s}\}$.

The proof of the second statement goes as follows. By Corollary 12.5.4 we can choose a relative interior point $\mathbf{p}$ of a $(d - 2)$-dimensional face F of $\mathbf{P}$. Let $\mathbf{q}_1$ and $\mathbf{q}_2$ be the centers of the generating balls of $\mathbf{P}$ such that $F := S^{d-1}(\mathbf{q}_1, 1) \cap S^{d-1}(\mathbf{q}_2, 1) \cap \mathbf{P}$. Clearly, $\mathbf{p} \notin \operatorname{conv}_s((\mathbf{B}^d[\mathbf{q}_1, 1] \cap \mathbf{B}^d[\mathbf{q}_2, 1]) \setminus \{\mathbf{p}\}) \supset \operatorname{conv}_s(\mathbf{P} \setminus \{\mathbf{p}\})$. $\quad\square$

Corollary 12.5.6 *(Euler–Poincaré Formula)* *If* **P** *is an arbitrary standard d-dimensional ball-polyhedron, then*

$$1 + (-1)^{d+1} = \sum_{i=0}^{d-1} (-1)^i f_i(\mathbf{P}),$$

where $f_i(\mathbf{P})$ *denotes the number of i-dimensional faces of* **P**.

Proof: It follows from Theorem 12.5.1 and the fact that a ball-polyhedron in $\mathbb{E}^d$ is a convex body, hence its boundary is homeomorphic to $S^{d-1}(\mathbf{o}, 1)$. $\square$

References

1. P. K. Agarwal and J. Pach, Combinatorial Geometry, *John Wiley and Sons, New York*, 1995.
2. M. Aigner and G. M. Ziegler, Proofs from the Book, *Springer, Berlin*, 4th edition, 2010.
3. R. Alexander, A problem about lines and ovals, *Amer. Math. Monthly* **75** (1968), 482–487.
4. R. Alexander, Lipschitzian mappings and the total mean curvature of polyhedral surfaces (I), *Trans. Amer. Math. Soc.* **288/2** (1985), 661–678.
5. G. Ambrus and F. Fodor, A new lower bound on the surface area of a Voronoi polyhedron, *Periodica Math. Hungar.* **53/1-2** (2006), 45–58.
6. E. M. Andreev, On convex polyhedra in Lobachevski spaces, *Mat. Sb. (N.S.)* **81** (1970), 445–478.
7. E. M. Andreev, On convex polyhedra of finite volume in Lobachevski spaces, *Mat. Sb. (N.S.)* **83** (1970), 256–260.
8. K. Anstreicher, The thirteen spheres: A new proof, *Discrete Comput. Geom.* **31** (2004), 613–625.
9. C. Bachoc and F. Vallentin, New upper bounds for kissing numbers from semidefinite programming, *J. Amer. Math. Soc.* **21** (2008), 909–924.
10. C. Bachoc and F. Vallentin, Semidefinite programming, multivariate orthogonal polynomials, and codes in spherical caps, *European J. Combin.* **30** (2009), 625–637.
11. K. Ball and A. Pajor, Convex bodies with few faces, *Proc. Am. Math. Soc.* **110** (1990), 225–231.
12. K. Ball, The plank problem for symmetric bodies, *Invent. Math.* **104** (1991), 535–543.
13. K. Ball, A lower bound for the optimal density of lattice packings, *Internat. Math. Res. Notices* **10** (1992), 217–221.
14. K. Ball, An elementary introduction to modern convex geometry, in Flavors of Geometry, S. Levy (Ed.), *Cambridge University Press. Math. Sci. Res. Inst. Publ.* **31**, 1997, 1–58.
15. W. Banaszczyk, Inequalities for convex bodies and polar reciprocal lattices in R^n II: Application of K-convexity, *Discrete Comput. Geom.* **16** (1996), 305–311.

16. W. Banaszczyk, A. E. Litvak, A. Pajor, and S. J. Szarek, The flatness theorem for nonsymmetric convex bodies via the local theory of Banach spaces, *Math. Oper. Res.* **24/3** (1999), 728–750.

17. T. Bang, On covering by parallel-strips, *Mat. Tidsskr. B.* **1950** (1950), 49–53.

18. T. Bang, A solution of the "Plank problem", *Proc. Am. Math. Soc.* **2** (1951), 990–993.

19. E. Bannai and N. J. A. Sloane, Uniqueness of certain spherical codes, *Canad. J. Math.* **33** (1981), 437–449.

20. E. Baranovskii, On packing n-dimensional Euclidean space by equal spheres, *Iz. Vissih Uceb. Zav. Mat.* **39/2** (1964), 14–24.

21. A. Barg and O. R. Musin, Codes in spherical caps, *Adv. Math. Comm.* **1** (2007), 131-149.

22. A. Barvinok, A Course in Convexity, *Graduate Studies in Mathematics, Amer. Math. Soc.* **54**, 2002.

23. I. Bárány and N. P. Dolbilin, A stability property of the densest circle packing, *Monatsh. Math.* **106** (1988), 107–114.

24. M. Belk and R. Connelly, Making contractions continuous: a problem related to the Kneser–Poulsen conjecture, *Contrib. to Discrete Math.* (to appear), 1–10.

25. V. Benci and F. Giannoni, Periodic bounce trajectories with a low number of bounce points, *Ann. Inst. H. Poincar Anal. Non Linaire* **6/1** (1989), 73–93.

26. M. Bern and A. Sahai, Pushing disks together - the continuous-motion case, *Discrete Comput. Geom.* **20** (1998), 499–514.

27. U. Betke and M. Henk, Finite packings of spheres, *Discrete Comput. Geom.* **19** (1998), 197–227.

28. U. Betke, M. Henk, and J. M. Wills, Finite and infinite packings, *J. reine angew. Math.* **53** (1994), 165–191.

29. U. Betke, M. Henk, and J. M. Wills, Sausages are good packings, *Discrete Comput. Geom.* **13** (1995), 297–311.

30. A. Bezdek, Solid packing of circles in the hyperbolic plane, *Studia Sci. Math. Hung.* **14** (1979), 203–207.

31. A. Bezdek and K. Bezdek, A note on the ten-neighbour packings of equal balls, *Beiträge zur Alg. und Geom.* **27** (1988), 49–53.

32. A. Bezdek and K. Bezdek, A solution of Conway's fried potato problem, *Bull. London Math. Soc.* **27/5** (1995), 492–496.

33. A. Bezdek and K. Bezdek, Conway's fried potato problem revisited, *Arch. Math.* **66/6** (1996), 522–528.

34. A. Bezdek, K. Bezdek, and R. Connelly, Finite and uniform stability of sphere packings, *Discrete Comput. Geom.* **20** (1998), 111–130.

35. A. Bezdek, On a generalization of Tarski's plank problem, *Discrete Comput. Geom.* **38** (2007), 189–200.

36. D. Bezdek, Dürer's unsolved geometry problem, *Canada-Wide Science Fair, St. John's* (May 15–23, 2004), 1–42.

37. D. Bezdek, A proof of an extension of the icosahedral conjecture of Steiner for generalized deltahedra, *Contrib. to Discrete Math.* **2/1** (2007), 86–92.

38. D. Bezdek and K. Bezdek, Shortest billiard trajectories, *Geom. Dedicata* **141** (2009), 197–206.

39. K. Bezdek, Ausfüllung eines Kreises durch kongruente Kreise in der hyperbolischen Ebene, *Studia Sci. Math. Hung.* **17** (1982), 353–366.

40. K. Bezdek, Circle-packings into convex domains of the Euclidean and hyperbolic plane and the sphere, *Geometriae Dedicata* **21** (1986), 249–255.

41. K. Bezdek and R. Connelly, Intersection points, *Ann. Univ. Sci. Budapest Sect. Math.* **31** (1988), 115–127.
42. K. Bezdek and R. Connelly, Covering curves by translates of a convex set, *Amer. Math. Monthly* **96/9** (1989), 789–806.
43. K. Bezdek, The problem of illumination of the boundary of a convex body by affine subspaces, *Mathematika* **38** (1991), 362–375.
44. K. Bezdek, Hadwiger's covering conjecture and its relatives, *Amer. Math. Monthly* **99** (1992), 954–956.
45. K. Bezdek, On the illumination of smooth convex bodies, *Arch. Math.* **58** (1992), 611–614.
46. K. Bezdek, Research problem 46, *Period. Math. Hungar.* **24** (1992), 119–121.
47. K. Bezdek, Hadwiger–Levi's covering problem revisited, in New Trends in Discrete and Computational Geometry, J. Pach (Ed.), *Springer, New York*, 1993.
48. K. Bezdek, A note on the illumination of convex bodies, *Geometriae Dedicata* **45** (1993), 89–91.
49. K. Bezdek, Gy. Kiss, and M. Mollard, An illumination problem for zonoids, *Israel J. Math.* **81** (1993), 265–272.
50. K. Bezdek, On affine subspaces that illuminate a convex set, *Beiträge zur Alg. und Geom.* **35** (1994), 131–139.
51. K. Bezdek and T. Hausel, On the number of lattice hyperplanes which are needed to cover the lattice points of a convex body, in Intuitive Geometry, *Coll. Math. Soc. J. Bolyai* **63**, *North-Holland, Amsterdam*, 1994, 27–31.
52. K. Bezdek and T. Zamfirescu, A characterization of 3-dimensional convex sets with an infinite X-ray number, in Intuitive Geometry, *Coll. Math. Soc. J. Bolyai* **63**, *North-Holland, Amsterdam*, 1994, 33–38.
53. K. Bezdek and T. Bisztriczky, A proof of Hadwiger's covering conjecture for dual cyclic polytopes, *Geom. Dedicata* **68** (1997), 29–41.
54. K. Bezdek, E. Daróczy-Kiss, and K. J. Liu, Voronoi polyhedra of unit ball packings with small surface area, *Periodica Math. Hungar.* **39/1–3** (1999), 107–118.
55. K. Bezdek, On a stronger form of Rogers' lemma and the minimum surface area of Voronoi cells in unit ball packings, *J. reine angew. Math.* **518** (2000), 131–143.
56. K. Bezdek, Improving Rogers' upper bound for the density of unit ball packings via estimating the surface area of Voronoi cells from below in Euclidean d-space for all $d \geq 8$, *Discrete Comput. Geom.* **28** (2002), 75–106.
57. K. Bezdek, On the maximum number of touching pairs in a finite packing of translates of a convex body, *J. Combin. Theory Ser. A* **98** (2002), 192–200.
58. K. Bezdek and R. Connelly, Pushing disks apart - the Kneser–Poulsen conjecture in the plane, *J. reine angew. Math.* **553** (2002), 221–236.
59. K. Bezdek, M. Naszódi, and B. Visy, On the mth Petty numbers of normed spaces, in Discrete Geometry, A. Bezdek (Ed.), *Marcel Dekker, New York*, 2003, 291–304.
60. K. Bezdek and P. Brass, On k^+-neighbour packings and one-sided Hadwiger configurations, *Beiträge zur Alg. und Geom.* **44** (2003), 493–498.
61. K. Bezdek and R. Connelly, The Kneser–Poulsen conjecture for spherical polytopes, *Discrete Comput. Geom.* **32** (2004), 101–106.
62. K. Bezdek, Sphere packings in 3-space, in *Proc. COE Workshop, Kyushu University, Fukuoka, Japan*, 2004, 32–49.

63. K. Bezdek and E. Daróczy-Kiss, Finding the best face on a Voronoi polyhedron - the strong dodecahedral conjecture revisited, *Monatshefte für Math.* **145** (2005), 191–206.

64. K. Bezdek, On the monotonicity of the volume of hyperbolic convex polyhedra, *Beiträge Algebra Geom.* **46/2** (2005), 609–614.

65. K. Bezdek, R. Connelly, and B. Csikós, On the perimeter of the intersection of congruent disks, *Beiträge Algebra Geom.* **47/1** (2006), 53–62.

66. K. Bezdek and M. Naszódi, Rigidity of ball-polyhedra in Euclidean 3-space, *European J. Combin.* **27** (2006), 255–268.

67. K. Bezdek, K. Böröczky, and Gy. Kiss, On the successive illumination parameters of convex bodies, *Periodica Math. Hungar.* **53/1-2** (2006), 71–82.

68. K. Bezdek, R. Connelly, and B. Csikós, On the perimeter of the intersection of congruent disks, *Beiträge Algebra Geom.* **47/1** (2006), 53–62.

69. K. Bezdek, Zs. Lángi, M. Naszódi, and P. Papez, Ball-polyhedra, *Discrete Comput. Geom.* **38/2** (2007), 201–230.

70. K. Bezdek and A. Litvak, On the vertex index of convex bodies, *Adv. Math.* **215/2** (2007), 626–641.

71. K. Bezdek and A. E. Litvak, Covering convex bodies by cylinders and lattice points by flats, *J. Geom. Anal.* **19/2** (2009), 233–243.

72. K. Bezdek and Gy. Kiss, On the X-ray number of almost smooth convex bodies and of convex bodies of constant width, *Canadian Math. Bull.* **52/3** (2009), 342–348.

73. K. Bezdek, Tarski's plank problem revisited, *arXiv:0903.4637v1* [math.MG] (March 26, 2009), 1–19.

74. K. Bezdek and R. Schneider, Covering large balls with convex sets in spherical space, *Beiträge Algebra Geom.* **51/1** (2010), 229–235.

75. M. Bezdek, On a generalization of the Blaschke–Lebesgue theorem for disk-polygons, *arXiv:0903.5361v1* [math.MG] (March 31, 2009), 1–11.

76. W. Blaschke, Konvexe Bereiche gegebener konstanter Breite und kleinsten Inhalts, *Math. Ann.* **76** (1915), 504–513.

77. B. Bollobás, Area of union of disks, *Elem. Math.* **23** (1968), 60–61.

78. V. Boltyanski, The problem of illuminating the boundary of a convex body, *Izv. Mold. Fil. AN SSSR* **76** (1960), 77–84.

79. V. G. Boltyanskii and M. Yaglom, Convex Figures, *Holt-Rinehart-Winston, New York,* 1961.

80. V. Boltyanski, A solution of the illumination problem for belt bodies, *Mat. Zametki* **58** (1996), 505–511.

81. V. Boltyanski, Solution of the illumination problem for three dimensional convex bodies, *Dokl. Akad. Nauk* **375** (2000), 298–301.

82. V. Boltyanski, Solution of the illumination problem for bodies with md M=2, *Discrete Comput. Geom.* **26** (2001), 527–541.

83. V. Boltyanski and H. Martini, Covering belt bodies by smaller homothetical copies, *Beiträge zur Alg. und Geom.* **42** (2001), 313–324.

84. V. Boltyanski, H. Martini, and P. S. Soltan, Excursions into Combinatorial Geometry, *Springer, New York,* 1997.

85. T. Bonnesen and W. Fenchel, Theory of Convex Bodies, *BCS Associates, Moscow, Id.,* 1987.

86. L. Bowen, Circle packing in the hyperbolic plane, *Math. Physics Electronic J.* **6** (2000), 1–10.

87. L. Bowen and C. Radin, Densest packing of equal spheres in hyperbolic space, *Discrete Comput. Geom.* **29** (2003), 23–39.

88. K. Böröczky, Gömbkitöltés állandó görbületű terekben, *Mat. Lapok* **25** (1974) (in Hungarian), 265–306.

89. K. Böröczky, Packing of spheres in spaces of constant curvature, *Acta Math. Acad. Sci. Hungar.* **32** (1978), 243–261.

90. K. Böröczky, The problem of Tammes for $n = 11$, *Studia Sci. Math. Hungar.* **18** (1983), 165–171.

91. K. Böröczky and K. Máthéné Bognár, Regular polyhedra and Hajós polyhedra, *Studia Sci. Math. Hungar.* **35** (1999), 415–426.

92. K. Böröczky, The Newton–Gregory problem revisited, in Discrete Geometry, A. Bezdek (Ed.), *Marcel Dekker, New York*, 2003, 103–110.

93. K. Böröczky and K. Böröczky Jr., Polytopes of minimal volume with respect to a shell - another characterization of the octahedron and the icosahedron, *Discrete Comput. Geom.* **38** (2007), 231–241.

94. P. Brass, On equilateral simplices in normed spaces, *Beiträge zur Alg. Geom.* **40** (1990), 303–307.

95. P. Brass, Erdős distance problems in normed spaces, *Comput. Geometry* **6** (1996), 195–214.

96. P. Brass, W. O. J. Moser and J. Pach, Research Problems in Discrete Geometry, *Springer, New York*, 2nd printing, 2006.

97. Y. D. Burago and V. A. Zalgaller, Geometric Inequalities, *Springer, New York*, 1988.

98. V. Capoyleas and J. Pach, On the perimeter of a point set in the plane, *DIMACS Ser., Discrete Math. and Th. Computer Sci., AMS, Providence, RI*, 1991, 67–76.

99. V. Capoyleas, On the area of the intersection of disks in the plane, *Comput. Geom.* **6/6** (1996), 393–396.

100. B. Casselman, The difficulties of kissing in three dimensions, *Notices of the AMS* **51/8** (2004), 884–885.

101. G. D. Chakerian, Sets of constant width, *Pacific J. Math.* **19** (1966), 13–21.

102. G. D. Chakerian, Inequalities for the difference body of a convex body, *Proc. Amer. Math. Soc.* **18** (1967), 879–884.

103. H. Cohn and N. Elkies, New upper bounds on sphere packings I, *Ann. Math.* **157** (2003), 689–714.

104. H. Cohn and A. Kumar, The densest lattice in twenty-four dimensions, *Electron. Res. Announc. Amer. Math. Soc.* **10** (2004), 58–67.

105. R. Connelly, The rigidity of certain cabled frameworks and the second order rigidity of arbitrarily triangulated convex surfaces, *Adv. Math.* **37/3** (1980), 272–299.

106. R. Connelly, Rigidity, in Handbook of Convex Geometry, *North Holland, Amsterdam*, 1993, 223–271.

107. R. Connelly and W. Whiteley, Second-order rigidity and pre-stress stability for tensegrity frameworks, *SIAM J. Discrete Math.* **9** (1996), 453–491.

108. J. H. Conway and N. J. A. Sloane, Sphere packings, lattices and groups, *Springer, New York*, 1999.

109. C. E. Corzatt, Covering convex sets of lattice points with straight lines, *Proceedings of the Sundance conference on combinatorics and related topics (Sundance, Ut., 1985), Congr. Numer.* **50** (1985), 129–135.

110. H. S. M. Coxeter, The polytopes with regular-prismatic vertex figures II, *Proc. London Math. Soc.* **34** (1932), 126–189.

111. H. S. M. Coxeter, An upper bound for the number of equal nonoverlapping spheres that can touch another of the same size, *Proc. Sympos. Pure Math.* **7** (1963), 53–71.

112. B. Csikós, On the Hadwiger–Kneser–Poulsen conjecture, *Bolyai Mathematical Studies Ser., Intuitive Geometry* **6** (1995), 291–300.

113. B. Csikós, On the volume of the union of balls, *Discrete Comput. Geom.* **20** (1998), 449–461.

114. B. Csikós, On the volume of flowers in space forms, *Geom. Dedicata* **86** (2001), 59–79.

115. B. Csikós, A Schläfli-type formula for polytopes with curved faces and its application to the Kneser–Poulsen conjecture, *Monatsh. Math.* **147** (2006), 273–292.

116. L. Danzer and B. Grünbaum, Über zwei Probleme bezüglich konvexer Körper von P. Erdős and von V.L. Klee, *Math. Zeitschrift* **79** (1962), 95–99.

117. L. Danzer, Finite point sets on S^2 with minimum distance as large as possible, *Discrete Math.* **60** (1986), 3–66.

118. H. Davenport and G. Hajós, Problem 35, *Mat. Lapok* **III/1** (1952), 94–95.

119. B. V. Dekster, The Jung theorem for spherical and hyperbolic spaces, *Acta Math. Hungar.* **67/4** (1995), 315–331.

120. B. V. Dekster, Each convex body in E^3 symmetric about a plane can be illuminated by 8 directions, *J. Geom.* **69/1-2** (2000), 37–50.

121. E. D. Demaine and J. O'Rourke, Geometric Folding Algorithms, *Cambridge University Press, Cambridge, UK*, 2007.

122. I. Dumer, Covering spheres with spheres, *Discrete Comput. Geom.* **38** (2007), 665–679.

123. Y. Edel, E. M. Rains and N. J. A. Sloane, On kissing numbers in dimensions 32 to 128, *Electronic J. of Comb.* **5** (1998), R22.

124. H. Edelsbrunner, The union of balls and its dual shape, *Discrete Comput. Geom.* **13/3-4** (1995), 415–440.

125. H. G. Eggleston, Sets of constant width in finite dimensional Banach spaces, *Israel J. Math.* **3** (1965), 163–172.

126. P. Erdős and C. A. Rogers, Covering space with convex bodies, *Acta Arith.* **7** (1962), 281–285.

127. P. Erdős and C. A. Rogers, The star number of coverings of space with convex bodies, *Acta Arith.* **9** (1964), 41–45.

128. M. Farber and S. Tabachnikov, Topology of cyclic configuration spaces and periodic trajectories of multi-dimensional billiards, *Topology* **41/3** (2002), 553–589.

129. G. Fejes Tóth, Kreisüberdeckungen der Sphäre, *Studia Sci. Math. Hungar.* **4** (1969), 225–247.

130. G. Fejes Tóth, Ten-neighbor packing of equal balls, *Period. Math. Hungar.* **12** (1981), 125–127.

131. G. Fejes Tóth, A note on covering by convex bodies, *Canad. Math. Bull.* **52/3** (2009), 361-365.

132. L. Fejes Tóth, Über die dichteste Kugellagerung, *Math. Z.* **48** (1943), 676–684.

133. L. Fejes Tóth, On the densest packing of circles in a convex domain, *Norske Vid. Selsk. Fordhl., Trondheim* **21** (1948), 68–76.

134. L. Fejes Tóth, Lagerungen in der Ebene, auf der Kugel und im Raum, *Springer Verlag, Berlin-Göttingen-Heidelberg*, 1953.

135. L. Fejes Tóth, Regular Figures, *Pergamon Press, Tarrytown, NY*, 1964.

136. L. Fejes Tóth, Solid circle packings and circle coverings, *Studia Sci. Math. Hungar.* **3** (1968), 401–409.

137. L. Fejes Tóth, Remarks on a theorem of R. M. Robinson, *Studia Sci. Math. Hungar.* **4** (1969), 441–445.

138. L. Fejes Tóth, Research problem 13, *Period. Math. Hungar.* **6** (1975), 197–199.

139. L. Fejes Tóth and H. Sachs, Research problem 17, *Period. Math. Hungar.* **7** (1976), 125–127.

140. L. Fejes Tóth, Solid packing of circles in the hyperbolic plane, *Studia Sci. Math. Hungar.* **15** (1980), 299–302.

141. S. P. Ferguson, Sphere packings V, Pentahedral prisms, *Discrete Comput. Geom.* **36/1** (2006), 167–204.

142. P. W. Fowler and T. Tarnai, Transition from spherical circle packing to covering: geometrical analogues of chemical isomerization, *Proc. R. Soc. London* **452** (1996), 2043–2064.

143. Z. Füredi and J.-H. Kang, Covering the n-space by convex bodies and its chromatic number, *Discrete Math.* **308/19** (2008), 4495–4500.

144. H. Freudenthal and B. L. van der Waerden, On an assertion of Euclid, *Simon Stevin* **25** (1947), 115–121.

145. C. F. Gauss, Untersuchungen über die Eigenschaften der positiven ternären quadratischen formen von Ludwig August Seber, *J. reine angew. Math.* **20** (1840), 312–320.

146. M. Ghomi, Shortest periodic billiard trajectories in convex bodies, *Geom. Funct. Anal.* **14** (2004), 295–302.

147. E. D. Gluskin and A. E. Litvak, Asymmetry of convex polytopes and vertex index of symmetric convex bodies, *Discrete Comput. Geom.* **40/4** (2008), 528–536.

148. I. Gohberg and A. Markus, A problem on covering of convex figures by similar figures, *Izv. Mold. Fil. Akad. Nauk. SSSR* **10** (1960), 87–90.

149. I. Gorbovickis, Kneser–Poulsen conjecture for large radii, *Manuscript* (2009), 1-17.

150. M. Gromov, Monotonicity of the volume of intersections of balls, in Geometrical Aspects of Functional Analysis, *Springer Lecture Notes* **1267**, *Springer, New York*, 1987, 1–4.

151. H. Groemer, Abschätzungen für die Anzahl der konvexen Körper, die einen konvexen Körper berühren, *Monatsh. Math.* **65** (1961), 74–81.

152. P. M. Gruber and C. G. Lekkerkerker, Geometry of Numbers, *North-Holland, Amsterdam*, 1987.

153. P. M. Gruber, Convex and Discrete Geometry, *Grundlehren der mathematischen Wissenschaften* **336**, *Springer, Berlin*, 2007.

154. H. Hadwiger, Über Treffenzahlen bei translations gleichen Eikörpern, *Arch. Math.* **8** (1957), 212–213.

155. H. Hadwiger, Ungelöste Probleme, Nr. 38, *Elem. Math.* **15** (1960), 130–131.

156. T. C. Hales, Sphere packings I, *Discrete Comput. Geom.* **17** (1997), 1–51.

157. T. C. Hales, Sphere packings II, *Discrete Comput. Geom.* **18** (1997), 135–149.

158. T. C. Hales, A proof of the Kepler conjecture, *Ann. Math.* **162/2–3** (2005), 1065–1185.

159. T. C. Hales, Historical overview of the Kepler conjecture, *Discrete Comput. Geom.* **36/1** (2006), 5–20.

160. T. C. Hales and S.P. Ferguson, A formulation of the Kepler conjecture, *Discrete Comput. Geom.* **36/1** (2006), 21–69.

161. T. C. Hales, Sphere packings III, Extremal cases, *Discrete Comput. Geom.* **36/1** (2006), 71–110.

162. T. C. Hales, Sphere packings IV, Detailed bounds, *Discrete Comput. Geom.* **36/1** (2006), 111–166.

163. T. C. Hales, Sphere packings VI, Tame graphs and linear programs, *Discrete Comput. Geom.* **36/1** (2006), 205–265.

164. T. C. Hales and S. McLaughlin, A proof of the dodecahedral conjecture, *arXiv:math.MG/9811079* (1998) (with updates and improvements from 2007), 1–90.

165. T. C. Hales and S. McLaughlin, The dodecahedral conjecture, *J. Amer. Math. Soc.* (to appear), 1–46.

166. H. Harborth, Lösung zu Problem 664A, *Elem. Math.* **29** (1974), 14–15.

167. E. M. Harrell, A direct proof of a theorem of Blaschke and Lebesgue, *J. Geom. Anal.* **12/1** (2002), 81–88.

168. D. Hilbert, Mathematical problems, *Bull. Amer. Math. Soc.* **8** (1902), 437–479.

169. M. E. Houle, Theorems on the existence of separating surfaces, *Discrete Comput. Geom.* **6** (1991), 49–56.

170. W.-Y. Hsiang, On the sphere packing problem and the proof of Kepler's conjecture, *Int. J. Math.* 4/5 (1993), 739–831.

171. W.-Y. Hsiang, Least Action Principle of Crystal Formation of Dense Packing Type and Kepler's Conjecture, *World Sci., Singapore,* 2001.

172. F. John, Extremum problems with inequalities as subsidiary conditions, in Studies and Essays Presented to R. Courant on his 60th Birthday, *Interscience, New York,* 1948, 187–204.

173. G. A. Kabatiansky and V.I. Levenshtein, Bounds for packings on a sphere and in space, *Problemy Peredachi Informatsii* **14** (1978), 3–25.

174. V. Kadets, Weak cluster points of a sequence and coverings by cylinders, *Mat. Fiz. Anal. Geom.* **11/2** (2004), 161–168.

175. V. Kadets, Coverings by convex bodies and inscribed balls, *Proc. Amer. Math. Soc.* **133/5** (2005), 1491–1495.

176. J. Kahn and G. Kalai, A counterexample to Borsuk's conjecture, *Bull. Amer. Math. Soc. (N.S.)* **29/1** (1993), 60–62.

177. R. Kannan and L. Lovász, Covering minima and lattice-point-free convex bodies, *Ann. Math.* **128** (1988), 577–602.

178. G. Kertész, Nine points on the hemisphere, in Intuitive Geometry, *Coll. Math. Soc. J. Bolyai* **63**, *North-Holland, Amsterdam,* 1994, 189–196.

179. Gy. Kiss, Illumination problems and codes, *Period. Math. Hungar.* **39** (1999), 65–71.

180. Gy. Kiss and P.O. de Wet, Notes on the illumination parameters of convex polytopes, *Manuscript* (2009), 1–12.

181. V. Klee and V. S. Wagon, Old and new unsolved problems in plane geometry and number theory, *MAA Dolciani Mathematical Expositions* 1991.

182. H. Kneser, Der Simplexinhalt in der nichteuklidischen Geometrie *Deutsche Math.* **1** (1936), 337–340.

183. M. Kneser, Einige Bemerkungen über das Minkowskische Flächenmass, *Arch. Math.* **6** (1955), 382–390.

184. M. Lassak, Solution of Hadwiger's covering problem for centrally symmetric convex bodies in E^3, *J. London Math. Soc.* **30** (1984), 501–511.

185. M. Lassak, Covering the boundary of a convex set by tiles, *Proc. Amer. Math. Soc.* **104** (1988), 269–272.

186. M. Lassak, Illumination of three-dimensional convex bodies of constant width, *Proc. 4th Internat. Congress of Geometry, Aristotle Univ. of Thessaloniki,* 1997, 246–250.

187. D. G. Larman and C. Zong, On the kissing numbers of some special convex bodies, *Discrete Comput. Geom.* **21** (1999), 233–242.

188. H. Lebesgue, Sur le problemedes isoperimetres at sur les domaines de larguer constante, *Bull. Soc. Math. France C. R.* **7** (1914), 72–76.

189. J. Leech, The problem of thirteen spheres, *Math. Gazette* **41** (1956), 22–23.

190. V. I. Levenshtein, On bounds for packings in n-dimensional Euclidean space, *Dokl. Akad. Nauk SSSR* **245** (1979), 1299–1303.

191. J.H. Lindsey, Sphere packing in $\mathbb{R}^3$, *Mathematika*, **33** (1986), 417–421.

192. H. Maehara, Isoperimetric theorem for spherical polygons and the problem of 13 spheres, *Ryukyu Math. J.* **14** (2001), 41–57.

193. H. Maehara, The problem of thirteen spheres - a proof for undergraduates *European J. Combin.* **28/6** (2007), 1770–1778.

194. H. Martini, Some results and problems around zonotopes, *Coll. Math. Soc. J. Bolyai, Intuitive Geometry, Sioófok 1985, North Holland* **48** (1987), 383–418.

195. H. Martini and V. Soltan, Combinatorial problems on the illumination of convex bodies, *Aequationes Math.* **57** (1999), 121–152.

196. E. Meissner, Über Punktmengen konstanter Breite, *Vjschr. naturforsch. Ges. Zürich*, **56** (1911), 42–50.

197. J. Milnor, The Sclāfli differential equality, in Collected Papers, Vol. 1, *Publish or Perish, Houston*, 1994.

198. H. D. Mittelmann and F. Vallentin, High accuracy semidefinite programming bounds for kissing numbers, *arXiv:0902.1105v3* [math.OC] (June 26, 2009), 1–7.

199. J. Molnár, Ausfüllung und Überdeckung eines konvexen sphärischen Gebietes durch Kreise I, *Publ. Math. Debrecen* **2** (1952), 266–275.

200. J. Molnár, Kreislagerungen auf Flächen konstanter Krümmung, *Math. Annalen* **158** (1965), 365–376.

201. D. J. Muder, Putting the best face on a Voronoi polyhedron, *Proc. London Math. Soc.* **3/56** (1988), 329–348.

202. D. J. Muder, A new bound on the local density of sphere packings, *Discrete Comput. Geom.* **10** (1993), 351–375.

203. O. R. Musin, The problem of the twenty-five spheres, *Russian Math. Surv.* **58/4** (2003), 794–795.

204. O. R. Musin, The kissing number in four dimensions, *Ann. Math.* **168/1** (2008), 1–32.

205. O. R. Musin, The kissing problem in three dimensions, *Discrete Comput. Geom.* **35/3** (2006), 375–384.

206. O. R. Musin, The one-sided kissing number in four dimensions, *Periodica Math. Hungar.* **53/1–2** (2006), 209-225.

207. M. Naszódi, On a conjecture of Károly Bezdek and János Pach, *Periodica Math. Hungar.* **53/1–2** (2006), 227–230.

208. M. Naszódi and Zs. Lángi, On the Bezdek–Pach conjecture for centrally symmetric convex bodies, *Canad. Math. Bull.* **52/3** (2009), 407–415.

209. G. Nebe and N. J. A. Sloane, Table of densest packings presently known, *http://www.research.att.com/njas/lattices/density.html.*
210. G. Nebe and N. J. A. Sloane, Table of the highest kissing numbers presently known, *http://www.research.att.com/ njas/lattices/kiss.html.*
211. A. M. Odlyzko and N. J. A. Sloane, New bounds on the number of unit spheres that can touch a unit sphere in n dimensions, *J. Combin. Theory Ser. A* **26** (1979), 210–214.
212. I. Papadoperakis, An estimate for the problem of illumination of the boundary of a convex body in E^3, *Geom. Dedicata* **75** (1999), 275–285.
213. C. M. Petty, Equilateral sets in Minkowski spaces, *Proc. Amer. Math. Soc.* **29** (1971), 369–374.
214. F. Pfender and G. M. Ziegler, Kissing numbers, sphere packings and some unexpected proofs, *Notices AMS* **51/8** (2004), 873–883.
215. G. Pisier, The Volume of Convex Bodies and Banach Space Geometry, *Cambridge University Press, Cambridge, UK*, 1989.
216. E. T. Poulsen, Problem 10, *Math. Scand.* **2** (1954), 346.
217. C. A. Rogers, A note on coverings, *Mathematika* **4** (1957), 1–6.
218. C. A. Rogers, The packing of equal spheres, *J. London Math. Soc.* **3/8** (1958), 609–620.
219. C. A. Rogers, Packing and Covering, *Cambridge University Press, Cambridge, UK*, 1964.
220. C. A. Rogers and G. C. Shephard, The difference body of a convex body, *Arch. Math.* **8** (1957), 220–233.
221. C. A. Rogers and G. C. Shephard, Convex bodies associated with a given convex body, *J. London Math. Soc.* **33** (1958), 270–281.
222. B. Roth and W. Whiteley, Tensegrity frameworks, *Trans. Amer. Math. Soc.* **265/2** (1981), 419–446.
223. M. Rudelson, Contact points of convex bodies, *Israel J. Math.* **101** (1997), 93–124.
224. M. Rudelson, Random vectors in the isotropic position, *J. Funct. Anal.* **164** (1999), 60–72.
225. H. Sachs, No more than nine unit balls can touch a closed hemisphere, *Studia Sci. Math. Hungar.* **21** (1986), 203–206.
226. O. Schramm, Illuminating sets of constant width, *Mathematika* **35** (1988), 180–189.
227. O. Schramm, On the volume of sets having constant width, *Israel J. Math.* **63/2** (1988), 178–182.
228. O. Schramm, Rigidity of infinite (circle) packings, *J. Amer. Math. Soc.* **4/1** (1991), 127–149.
229. K. Schütte and B. L. van der Waerden, Das Problem der dreizehn Kugeln, *Math. Ann.* **125** (1953), 325–334.
230. R. Seidel, Exact upper bounds for the number of faces in d-dimensional Voronoi diagrams, *DIMACS Ser. Discrete Math. Th. Comput. Sci., Amer. Math. Soc., Appl. Geom. Discrete Math.* **4** (1991), 517–529.
231. P. S. Soltan, Analogues of regular simplices in normed spaces, *Dokl. Akad. Nauk. SSSR* **222** (1975), 1303–1305.
232. P. Soltan and V. Soltan, Illumination through convex bodies, *Dokl. Akad. Nauk. SSSR* **286** (1986), 50–53.
233. J. J. Stoker, Geometric problems concerning polyhedra in the large, *Com. Pure and Applied Math.* **21** (1968), 119–168.

234. V. N. Sudakov, Gaussian random processes and measures of solid angles in Hilbert space, *Dokl. Akad. Nauk. SSSR* **197** (1971), 43–45.

235. K. J. Swanepoel, Quantitative illumination of convex bodies and vertex degrees of geometric Steiner minimal trees, *Mathematika* **52/1-2** (2005), 47–52.

236. H. P. F. Swinnerton–Dyer, Extremal lattices of convex bodies, *Proc. Cambridge Philos. Soc.* **49** (1953), 161–162.

237. S. Tabachnikov, Geometry and billiards, *Amer. Math. Soc.* 2005.

238. I. Talata, Covering the lattice points of a convex body with affine subspaces, *Bolyai Soc. Math. Stud.* **6** (1997), 429–440.

239. I. Talata, Exponential lower bound for translative kissing numbers of d-dimensional convex bodies, *Discrete Comput. Geom.* **19** (1998), 447–455.

240. I. Talata, Solution of Hadwiger–Levi's covering problem for duals of cyclic 2k-polytopes, *Geom. Dedicata* **74** (1999), 61–71.

241. I. Talata, The translative kissing number of tetrahedra is 18, *Discrete Comput. Geom.* **22** (1999), 231–293.

242. W. P. Thurston, The Geometry and Topology of 3-Manifolds, *Princeton Univ. Lecture Notes, Princeton, NJ*, 1980.

243. N. Tomczak-Jaegermann, Banach-Mazur Distances and Finite-Dimensional Operator Ideals, *Pitman Monographs and Surveys in Pure and Applied Mathematics, John Wiley and Sons, New York*, 1989.

244. E.B. Vinberg, Discrete groups generated by reflections in Lobacevskii spaces, *Math. USSR-Sb.* **1** (1967), 429–444.

245. A. Wallace, Algebraic approximation of curves, *Canad. J. Math.* **10** (1958), 248–278.

246. B. Weissbach, Invariante Beleuchtung konvexer Körper, *Beiträge zur Alg. Geom.* **37** (1996), 9–15.

247. J. M. Wills, Finite sphere packings and sphere coverings, *Rend. Semin. Mat. Messina* **II/2** (1993), 91–97.

248. A. D. Wyner, Capabilities of bounded discrepancy decoding, *Bell Systems Tech. J.* **54** (1965), 1061–1122.

249. S. Zelditch, Spectral determination of analytic bi-axisymmetric plane domains, *Geom. Funct. Anal.* **10/3** (2000), 628–677.